From DNA to Diversity

From DNA to Diversity

*Molecular Genetics
and the Evolution of Animal Design*

Sean B. Carroll

Jennifer K. Grenier

Scott D. Weatherbee

Howard Hughes Medical Institute and
University of Wisconsin
Madison, Wisconsin

b

**Blackwell
Science**

© 2001 by Sean Carroll

Editorial Offices:
Commerce Place, 350 Main Street, Malden, Massachusetts 02148, USA
Osney Mead, Oxford OX2 0EL, England
25 John Street, London WC1N 2BL, England
23 Ainslie Place, Edinburgh EH3 6AJ, Scotland
54 University Street, Carlton, Victoria 3053, Australia

Other Editorial Offices:
Blackwell Wissenschafts-Verlag GmbH, Kurfürstendamm 57, 10707 Berlin, Germany
Blackwell Science KK, MG Kodenmacho Building, 7-10 Kodenmacho Nihombashi, Chuo-ku, Tokyo 104, Japan

Distributors:

USA   Blackwell Science, Inc.
       Commerce Place
       350 Main Street
       Malden, Massachusetts 02148
       (Telephone orders: 800-215-1000
       or 781-388-8250;
       fax orders: 781-388-8270)

Canada   Login Brothers Book Company
          324 Saulteaux Crescent
          Winnipeg, Manitoba R3J 3T2
          (Telephone orders: 204-837-2987)

Australia   Blackwell Science Pty, Ltd.
             54 University Street
             Carlton, Victoria 3053
             (Telephone orders: 03-9347-0300;
             fax orders: 03-9349-3016)

Outside North America and Australia
Blackwell Science, Ltd.
c/o Marston Book Services, Ltd.
P.O. Box 269
Abingdon, Oxon OX14 4YN
England
(Telephone orders: 44-01235-465500;
fax orders: 44-01235-465555)

Acquisitions: Nancy Whilton
Development: Jill Connor
Production: Irene Herlihy
Manufacturing: Lisa Flanagan
Marketing Manager: Carla Daves

Cover design by Leslie Haimes and Jamie Carroll
Interior design by Leslie Haimes
Typeset by D&G Limited, LLC

Printed and bound by Walsworth Publishing Company

Printed in the United States of America
00 01 02 03 5 4 3 2 1

The Blackwell Science logo is a trade mark of Blackwell Science Ltd., registered at the United Kingdom Trade Marks Registry

Library of Congress Cataloging-in-Publication Data

Carroll, Sean B.
    From DNA to diversity: molecular genetics and the evolution of animal design/by
    Sean B. Carroll, Jennifer K. Grenier, Scott D. Weatherbee.
        p. cm.
    Includes bibliographical references and index.
    ISBN 0-632-04511-6
    1. Evolutionary genetics. 2. Biological diversity. 3. Morphology (Animals) I. Grenier,
Jennifer K. II. Weatherbee, Scott D. III. Title.

QH390 .C37 2001
572.8'38--dc21

                                                                    00-062165

To our families

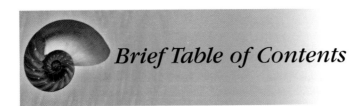

Brief Table of Contents

Chapter 1   A Brief History of Animals   1

Chapter 2   The Genetic Toolkit for Development   15

Chapter 3   Building Animals   51

Chapter 4   Evolution of the Toolkit   97

Chapter 5   Diversification of Body Plans and Body Parts   123

Chapter 6   The Evolution of Morphological Novelties   149

Chapter 7   From DNA to Diversity: The Primacy of Regulatory Evolution   173

Glossary   197

Index   203

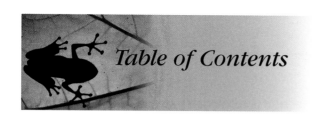

Table of Contents

PREFACE xiii

CHAPTER 1   **A Brief History of Animals**..1
             Animal Origins and the Fossil Record 2
             The Animal Tree 6
             General Features of Animal Design and Diversity...................8
             Evolution and Development: DNA and Diversity..................13
             Selected Readings 13

CHAPTER 2   **The Genetic Toolkit for Development**..................................15
             Before the Toolkit—Organizers, Fields, and Morphogens 15
             The Genetic Toolkit...18
                 The *Drosophila* Toolkit..19
                 Classifying genes according to their developmental function..............19
                 Homeotic genes and segmental identity 19
                 The homeobox 26
                 Field-specific selector genes..26
                 Compartment selector genes 28
                 Cell-type-specific selector genes...31
                 Formation of the body axes 32
                 Systematic searches for developmental genes in Drosophila 32
                 The anteroposterior axis 33
                 The dorsoventral axis...35
                 Expression of toolkit genes..35
                 Toolkit gene products: transcription factors and signaling
                    pathway components...36
                 Pleiotropy of toolkit genes 39
             Sharing of the Genetic Toolkit Among Animals 43
                 Hox genes..43
                 Field- and cell-type-specific selector genes 44
                 Signaling pathways: classical organizers and morphogens 46
             The Toolkit and Animal Design...47
             Selected Readings 48

CHAPTER 3   **Building Animals** 51
             Gene Regulation in Metazoans 51
             The Architecture of Genetic Regulatory Hierarchies 53
                 General features and approach..53
                 Regulatory logic—pathways, circuits, batteries, feedback loops,
                    networks, and the connectivity of genes 55

Model regulatory hierarchies and the key genetic switches that
operate them 57
The Insect Body Plan 57
From egg to segments: the anteroposterior coordinate system 58
Generation of maternal transcription factor gradients 60
Transcriptional activation of and cross-regulation by gap genes 60
Initiation of periodic pair-rule gene expression 61
Regulation of segment polarity genes by pair-rule proteins 62
General lessons from the segmentation genes....................................62
The dorsoventral axis coordinate system 65
The *Hox* ground plan 65
Secondary fields: integrating the anteroposterior and dorsoventral
coordinate systems..68
The limb fields 69
The wing primordia 69
The salivary gland 69
Neural and muscle precursors...70
Patterning within secondary fields: organizing signals and
selector genes..71
The anteroposterior coordinate system..72
The dorsoventral coordinate system 72
Signal integration by the vestigial field-specific selector gene...............72
Combinatorial regulation of wing patterning by signaling proteins
and the Vg/Sd selector proteins...74
The response to organizing signals 74
The action of the Vg/Sd selector genes...74
The modification of regulatory hierarchies in secondary fields by
Hox genes...75
The Vertebrate Body Plan 77
The anteroposterior axis: somite formation and segmentation............77
The vertebrate *Hox* ground plan...80
Vertebrate limb development 82
Positioning of the limb buds along the rostrocaudal axis 84
Outgrowth of the limb bud: dorsoventral and
anteroposterior regulatory hierarchies...84
Integration of organizing signals to form the proximodistal axis
and deployment of *Hox* genes in the limb field 87
Regulatory networks controlling the differentiation of major
limb pattern elements..89
The regulation of forelimb and hindlimb identity by selector
genes 89
Review: The General Logic and Mechanisms Controlling Gene
Expression in Cellular Fields..90
Selected Readings 92

CHAPTER 4   **Evolution of the Toolkit** **97**

The History of Gene Families 98
    Conservation of developmental regulatory genes: phylogenetic
       inferences about animal ancestors..98
    Gene duplication...101
    Mapping gene duplication events onto an animal phylogeny...........104
    Gene divergence 105
    Assembly of the toolkit: the first animals 105
Case Study: Evolution of the *Hox* Complex 108
    Expansion of the *Hox* complex...108
    The *ParaHox* genes: a sister complex to the *Hox* genes..................110
    New functions for some insect *Hox* genes 111
Interpreting the Toolkit: Inferences About Animal Evolution 112
    Expansion of the toolkit and the evolution of morphological
       complexity...112
    Conserved genes, conserved biochemical functions 113
    Conserved developmental functions: rebuilding the bilaterian
       ancestor from phylogenetic inference 114
The Toolkit as Developmental Potential 117
Selected Readings 119

CHAPTER 5   **Diversification of Body Plans and Body Parts** **123**

Diversity of Anterior/Posterior Body Organization within
    Arthropods and Vertebrates..124
    Evolution of the genetic control of segmentation in arthropods 125
    Shifts in trunk *Hox* gene expression that mirror changes in
       arthropod body architecture 127
    Hox genes and the evolution of arthropod heads 130
    Annelid *Hox* expression patterns 131
    Correlation of vertebrate axial patterning with *Hox* expression
       domains 131
    How do *Hox* domains shift during evolution? 133
Morphological Diversity Within a Conserved Body Plan 135
    Evolution of the limbless insect abdomen 135
    Evolution of insect wing number..136
    Diversification of insect wing morphology 139
    Modulations in *Hox* expression patterns within fields that
       contribute to insect diversity 140
    Vertebrate limb diversity: regulatory changes downstream of
       other selector genes...142
Regulatory Evolution and the Diversification of Homologous
    Body Parts 145
Selected Readings 146

CHAPTER 6   **The Evolution of Morphological Novelties** **149**
What Is Morphological Novelty? 150
Novel Functions from Older Morphological Structures 150
    Epithelial appendages: scales, feathers, and hair...............................150
    The evolution of insect wings from dorsal leg structures 152
    Butterfly wing color scales: an evolutionary canvas.........................154
    Evolution of the butterfly eyespot 155
The Evolution of Vertebrate Novelties......................................159
    Fins to limbs: paired appendages and the tetrapod hand.................159
    Evolution of the neural crest and the vertebrate brain......................161
    Evolution of the notochord 163
Evolution of Radical Body Plan Changes 163
    Loss of the ascidian tail: a chordate lacking chordate features..........163
    Limbless tetrapods, or how the snake lost its legs 165
    Evolution of the echinoderm body plan 166
Regulatory Evolution and the Origin of Novelties..............................167
Selected Readings 169

CHAPTER 7   **From DNA to Diversity: The Primacy of Regulatory Evolution** ...**173**
Why Is Regulatory Evolution a Primary Force in Morphological
    Evolution? 173
The Function and Evolution of *Cis*-Regulatory DNA...........................175
    Functions of *cis*-regulatory elements 175
    Evolution of *cis*-regulatory elements 175
    De novo evolution of cis-*regulatory elements from preexisting*
        nonfunctional DNA sequences...176
    Evolution of cis-*regulatory elements from existing elements*...............180
    The evolution of gene repression versus gene activation.................181
    Case studies in *cis*-regulatory evolution 182
    Conservation of functional elements among widely
        divergent taxa 182
    The dynamics of sequence turnover in cis-*regulatory DNA*...............183
    Functional modification of cis-*regulatory DNA sequences*.................186
The Role of *Cis*-Regulatory DNA in Morphological Variation and
    the Response to Selection 186
    Cryptic genetic variation and the potential for morphological
        evolution...188
The Evolution of Regulatory DNA and Morphological Diversity..........190
    Extrapolating from bristles to body plans 190
Selected Readings 194

GLOSSARY 197

INDEX..203

Preface

The Earth is now populated by between 1 million and perhaps as many as 20 million animal species, which represent probably less than 1% of all animal species that have ever existed. An even more remarkable fact is that all of this diversity—aardvarks and ostriches, butterflies and pythons, dinosaurs and earthworms—descended from a common bilaterally symmetrical ancestor that lived in Precambrian seas more than 540 million years ago. Traditionally approached through paleontology, systematics, and comparative anatomy, the story of animal evolution has, until recently, been sorely missing one huge chapter—namely, genetics.

Animals diverge from common ancestors through changes in their DNA. The major question, then, is, Which changes in DNA account for morphological diversity? The answers to this question have eluded us for the half-century since the formulation of Modern Synthesis and the discovery of the structure of DNA. Although many reasons exist to explain this omission, foremost among them is that biology first had to address another central genetic mystery—that is, which genes out of the thousands in any species control morphology?

One of the most important biological discoveries of the past two decades is that most animals, no matter how divergent in form, share specific families of genes that regulate major aspects of body pattern. The discovery of this common genetic "toolkit" for animal development has had two major impacts. First, it has enabled biologists to uncover widely conserved molecular, cellular, and developmental processes whose existence was concealed by previously incomparable anatomies. Second, it has focused the study of the genetic basis of animal diversity on how the number, regulation, and function of genes within the toolkit have changed over the course of animal evolution.

The genetic picture of morphological diversity presented in this book is highly influenced by the legacy of previous successes of genetic logic. The mysteries of enzyme induction in bacteria and bacteriophage life cycles were, through formal genetic logic and molecular biology, ultimately reduced to elegant genetic switches that determined the on/off state of groups of genes. This success laid the foundation for understanding the regulation of genes in different cell types of multicellular organisms and, in turn, the regulation of genes in space and over time during the development of individual organisms. Similarly, recent advances in understanding how the toolkit operates in the design of just a few model species has laid the foundation for attempts to unravel the evolution of a wide variety of animal structures and patterns.

The presentation in this book lies at the intersection of evolutionary biology with embryology and genetics. Comprehensive treatment of any of these long-established, fast-growing disciplines can be found in full textbooks dedicated to each. Because our goal is to elucidate general principles about the genetic basis of morphological change, we will focus on those genes, developmental processes, and taxa that are best known and best illustrate these principles.

The book is organized into two parts. The first part (Chapters 1–3) focuses on the history of animals and on animal developmental genetics and regulatory mechanisms. We first examine some of the major trends in animal design and evolution illustrated in the fossil record and by modern forms (Chapter 1). Next, we take an inventory of the genetic toolkit for the development of model species (Chapter 2). Finally, we analyze the regulation and function of these genes in the complex hierarchies that govern animal development (Chapter 3). This crucial background knowledge of the major transitions in animal evolution and the genetic logic of animal design sets the stage for the analysis of mechanisms of morphological evolution.

The second part of the book (Chapters 4–7) examines the genetic mechanisms underlying the evolution of animals at different morphological levels. We take a case study approach by focusing on the best-understood examples of the evolution of the genetic toolkit (Chapter 4), the diversity of body plans and body parts (Chapter 5), and novel structures (Chapter 6). In the final chapter (Chapter 7), we discuss why and how changes in gene regulation have played a primary role in the evolution of diversity across the morphological spectrum—from small-scale differences within or between species, to the large-scale differences that distinguish higher taxa.

We have provided selected references for further reading at the end of each chapter. By no means should these citations (or this book) be taken as the primary or exclusive references on a topic. For both brevity and to circumvent questions of priority in ideas or evidence, we have avoided attributions to specific authors in the text.

One of the inspirations for our approach was Mark Ptashne's classic *A Genetic Switch,* in which many of the basic physiological and molecular principles of gene regulation were illuminated by focusing on the bacteriophage λ. In the preface, Ptashne stated that "one of the charms of molecular biology is that the answers it provides to fundamental questions for the most part can be easily visualized." Few fields in biology can rival the aesthetic appeal of the new comparative embryology. Indeed, the visualization of members of the genetic toolkit in action during the development of different species has already become a surrogate for analyzing final forms. For those who find conceptual beauty in the logic and molecular anatomy of genetic switches, the genetic switches controlling animal anatomy may be even more appealing. Not only do they control the striking patterns of gene expression within developing embryos, but, as we shall see, they are also key to understanding how the wonderful, but presently dwindling diversity of animal forms has evolved.

ACKNOWLEDGMENTS

We are indebted to many people who have made the writing and illustration of this book possible. Jamie Wilson Carroll endured innumerable drafts of every chapter and the progressively less intelligible scrawl of the authors in the editing process. Leanne Olds created most of the artwork—an enormous job that was always completed cheerfully and with amazing speed. Steve Paddock rendered many of the digital microscopy figures. Steve's taste, creativity, and stewardship of a large image data base are most appreciated. And many thanks to our colleagues who provided additional illustrations used throughout the book.

We are grateful to several individuals who reviewed the book as a whole or in part. Thanks to Craig Brunetti, Dave Lewis, Craig Nelson, John True, Artyom Kopp, Nicole King, Ron Galant, Trisha Wittkopp, and John Fallon. Special thanks to Georg Halder for tackling the whole manuscript. We also thank a large group of anonymous reviewers who provided many suggestions that helped to clarify the concepts and the text clearer. We hope that you are pleased with the results.

We are especially grateful to colleagues who have shared their expertise and ideas in collaborations with us over the past several years. Special thanks to Allen Laughon, Matt Scott, Paul Brakefield, Cliff Tabin, Neil Shubin, Andy Knoll, Fred Nijhout, Vernon French, Michael Akam, Andre Adoute, Paul Whitington, Steve Irvine, and Mark Martindale for their generous efforts and to Eric Davidson, Joram Piatigorsky, Nipam Patel, Craig Nelson, and Georg Halder for discussions on topics covered here.

This book was only made possible by the blood, sweat, and tears of many years of work by past and present members of the Carroll laboratory at the University of Wisconsin–Madison to whom we (especially Sean Carroll) owe an enormous debt. Thanks to Bruce Thalley, Stephanie Vavra, Jim Skeath, Jim Langeland, Nadean Brown, Dave Keys, Angela Hudson, Jeff Esch, Bob Warren, Ron Galant, Trisha Wittkopp, Elizabeth Jones, Teresa Orenic, Jim Williams, Lisa Nagy, Grace Panginiban, Georg Halder, Jaeseob Kim, Mary Ellen Kraus, Ted Garber, Dave Lewis, Craig Brunetti, John True, Kirsten Guss, Craig Nelson, Artyom Kopp, Steve Paddock, Nicole King, Jeff Magee, Kathy Vaccaro, Jayne Selegue, Vicky Kassner, John Schlitz, Julie Gates, Brian Rollman, Sherrie Neas, Bret Pearson, and Sarah Hutchinson. Work in the Carroll laboratory has been generously supported by the Howard Hughes Medical Institute, the National Science Foundation, the Shaw Scientists's Program of the Milwaukee Foundation, and the Human Frontiers Science Program.

CHAPTER 1

A Brief History of Animals

The central focus of this book is to identify the genetic mechanisms underlying the evolution of animal design. To approach this mystery, new discoveries and ideas from developmental genetics must be integrated into the larger framework of the evolutionary history of animal life. This history is reconstructed from many fields of study—in particular, paleontology, systematics, and comparative biology. In this chapter, we present a brief overview of animal evolution from these three perspectives. This discussion provides a historical foundation for the consideration of the mechanistic questions that are addressed in subsequent chapters.

First, we discuss the origin of animals and the radiation of the major animal phyla based on evidence gleaned from the fossil record. Most living phyla have ancient origins, and the fundamental differences between them evolved long ago. Two milestones in early animal history that are of special interest are the evolution of bilaterally symmetric animals and the explosive radiation of these forms in the Cambrian period more than 500 million years ago.

Second, we examine the phylogenetic relationships among animals. Understanding the direction of evolutionary change in morphological, developmental, or genetic traits and the ability to make inferences about animal ancestors requires knowledge of the structure of the animal evolutionary tree. New molecularly based phylogenies have revealed unexpected relationships among anatomically disparate phyla, refuting long-held notions from morphological comparisons about which animals are more closely related.

Third, we consider the comparative anatomy of selected phyla with the aim of identifying some of the major trends in the evolutionary diversification of individual phyla. In particular, we focus on the modular organization of the body plans and body parts of larger animals—the vertebrates, arthropods, and annelids. Much of the large-scale morphological diversity within these phyla (for example, between different classes) involves differences in the number and pattern of modular elements (segments, appendages, and so on). The recognition of the modular organization of these animals is an important conceptual link to understanding the genetic logic controlling their development and the mechanisms underlying the evolution of diversity.

"... an understanding of regulation must lie at the center of any rapprochement between molecular and evolutionary biology; for a synthesis of the two biologies will surely take place, if it occurs at all, on the common field of development."

—Stephen Jay Gould, Ontogeny and Phylogeny (1977)

ANIMAL ORIGINS AND THE FOSSIL RECORD

The fossil record is our primary window into the history of life. It provides many kinds of information that cannot be inferred from living animals. Fossils give us pictures of extinct forms that may be ancestors of modern animals, provide minimal estimates of the time of origin or divergence of particular groups, reveal episodes of extinctions and radiations, and, in favorable circumstances, offer detailed accounts of the evolution of important structures.

The search for the origins of modern animals begins with an assessment of the **Cambrian** fossil record. Since before Darwin's time, it has been known that animal diversity increased dramatically during this period, which spans an age from roughly 545 to 490 million years ago (Ma). Molluscs, arthropods, annelids, chordates, echinoderms, and representatives of most other modern phyla make their first appearance in Cambrian fossil deposits (Fig. 1.1).

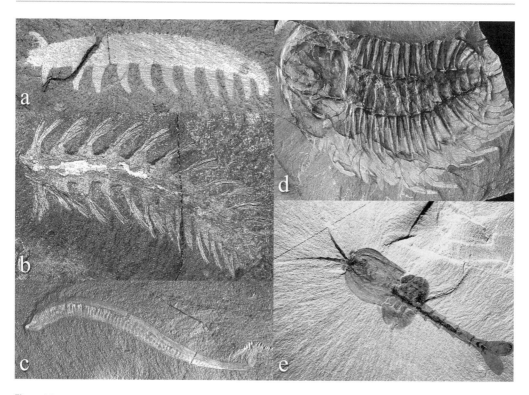

Figure 1.1
Cambrian animal fossils

Representatives of most modern phyla are found in Cambrian deposits and are made up of repeating units. (**a**) *Aysheaia pedunculata,* an onychophoran; (**b**) *Burgessochaeta setigera,* a polychaete annelid; (**c**) *Pikaia gracilens,* a chordate; (**d**) *Olenoides serratus,* a trilobit-omorph arthropod; and (**e**) *Waptia fieldensis,* a crustacean-type arthropod.

Source: Photographs from Briggs DEG, Erwin DH, Collier FJ. Fossils of the Burgess shale. Washington, DC: Smithsonian Institution Press, 1994; reprinted by permission from the Smithsonian Institution Press.

The emergence of large, complex animal forms and their radiation over a 10 to 25 million year interval in the Early–Middle Cambrian is often referred to as the "Cambrian Explosion."

The appearance of these animals in the Cambrian fossil record gives us only a minimum estimate of their time of origin. The crucial question about the Cambrian Explosion is whether it marks the origin of animals or the origin of modern phyla. Did most phyla evolve in this short period, or did they predate their preservation in the Cambrian fossil record? Although the Precambrian animal fossil record is relatively scarce, several kinds of fossil evidence indicate that the origins of most modern phyla predate the Cambrian. First, the fossil record of some modern groups clearly begins before this period. For example, body fossils of both cnidarians and sponges predate the Cambrian (Fig. 1.2). The Cnidaria are **diploblastic** animals, composed of two tissue layers, and have a radially symmetric body design that distinguishes them from sponges and from a much larger number of modern phyla that are **triploblastic**—that is, composed of three tissue layers—and have bilaterally symmetrical body designs (the **Bilateria**). Second, Precambrian deposits contain evidence in the form of **trace fossils**, the record of the meanderings and burrowings of animals in sediments, which indicate the existence of some bilaterian forms (see Figs. 1.2 and 1.3d) well before the Cambrian Explosion. A third piece of potential evidence for earlier animal origins is the **Ediacaran** fauna (575–544 Ma), named for the Australian locale in which they were first discovered.

The biological interpretation of Ediacaran fossils and their relationships, if any, to modern animals remains controversial. Several distinct body plans have been identified, including radially symmetric types and a number of frond-like and tube-like forms (see Fig. 1.3). None of these bear any clear-cut similarity to modern animals, so they have been difficult to place on the tree of animal evolution. Some of the Ediacaran fossils could represent diploblastic forms related to cnidarians or sponges. Others could be primitive bilaterians that possess some, but not all, features of modern bilaterians.

The difficulties in placing Ediacarans in the scheme of animal evolution have led to the proposal that they represent an extinct experiment in multicellular life. On the other hand, perhaps their lack of resemblance to modern groups is exactly what should be expected of primitive animals. It is possible that the Ediacaran fauna include both extinct types of diploblastic animals and primitive ancestors of modern bilaterians. The fossil record indicates that some Ediacaran forms persisted into the Cambrian, but then died out as bilaterians, sponges, cnidarians, and ctenophores flourished.

Given the uncertainty of the relationship of the Ediacarans to modern phyla and the paucity of body fossils prior to the Cambrian, it is difficult to pinpoint the origins of modern animals based on the fossil evidence. Consequently, biologists have turned to other methods to try to identify when major animal groups diverged. Using the evolution of protein and ribosomal RNA sequences between species to calibrate **molecular clocks**, estimates of the time of divergence of most animal phyla have been made that range from approximately 650 Ma to more than 1000 Ma. While these estimates remain controversial, even the most conservative estimate suggests a period of more than 100 million years before the beginning of the Cambrian in which most bilaterian phyla had evolved but led a paleontologically cryptic existence.

It is widely believed that primitive bilaterians may have been very small and their size limited by atmospheric and oceanic oxygen levels. This fact would help to explain their slim fossil record before the Cambrian (see Fig. 1.2). In the last few years, evidence has

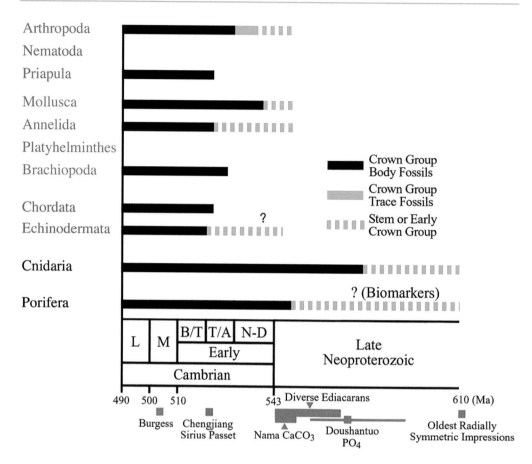

Figure 1.2
The early fossil record of animals

The appearance of various animal phyla are indicated, relative to the Cambrian and Proterozoic periods. The ages of fossils from particular localities are shown in red at the bottom right. Note that the cnidarian and poriferan records clearly predate the Cambrian. Other phyla first appear in the Cambrian, although early members may exist that predate the Cambrian by a considerable period.
Source: Adapted from Knoll AH, Carroll SB. Science 1999;284:2129–2137.

also been gathered that suggests a possible mass extinction at the boundary between the Proterozoic and Cambrian. Whatever the cause of such an event, it may have hastened the extinction of Ediacaran forms and opened up the ecological opportunity for bilaterians to radiate. Environmental and ecological changes may have removed constraints on bilaterians, permitting the evolution of larger animals. In addition, competitive interactions among bilaterians may have facilitated the evolution of skeletonized taxa, more sophisticated predatory and defense behaviors, and the variety of anatomical innovations that unfolded in the Cambrian.

Figure 1.3
Precambrian animal fossils and traces

(**a**) *Ediacaria,* a radially symmetrical form from deposits in Australia. (**b**) Calcified fossils in limestone from Namibia. (**c**) *Pteridinium,* a frond-like Ediacaran fossil form built of repeating units. (**d**) Trace fossils made in sediments by bilaterian animals.

Source: Knoll AH, Carroll SB. Science 1999;284:2129–2137.

THE ANIMAL TREE

There are about 35 living animal phyla. To understand the origin and evolution of any feature found in one or more of these groups, it is necessary to have a picture of the phylogenetic relationships among animals. Ideally, the fossil record would present a complete, ordered, unambiguous picture of the branching pattern of the animal tree. Unfortunately, it does not. As the divergence of most bilaterian phyla appears to have predated the emergence of recognizable members of modern phyla, we must make our inferences from later, more derived forms.

Constructing an accurate picture of metazoan relationships has been challenging, and many alternative schemes of animal phylogeny have been proposed and scrutinized over recent decades. Most approaches have relied on anatomical and embryological comparisons. Different trees emerge when different characters are used or when the same characters are weighted differently. In general, phyla are grouped according to shared characters. For example, all arthropods are segmented and possess jointed limbs. What is most difficult to determine is whether apparent similarities between animals (for example, segmentation in arthropods and annelids) are due to a close relationship, are superficial, or evolved independently.

One way to circumvent the reliance on morphological comparisons is to use molecular genetic characters to construct animal phylogenies. As taxa diverge, the sequences of DNA, RNA, and protein molecules diverge as well; the relative degree of divergence can therefore be used to assess phylogenetic relationships. In addition, the presence or absence of particular genes, or the linkage of a group of genes on chromosomes, can be used to construct phylogenetic trees. New molecularly based methods have been combined with morphology-based approaches to both prune and strengthen the animal tree.

We now recognize shared morphological, developmental, and genetic traits that suggest that the Bilateria can be organized into three great **clades** (a set of species descended from a common ancestor) (Fig. 1.4):

- The **deuterostomes**, including chordates, echinoderms, ascidians, and hemichordates. The deuterostomes are named for a shared feature of early embryonic development in which the mouth forms from a site separate from the **blastopore**, an opening in the early embryo.

- Two groups of **protostomes**, in which the mouth develops from the blastopore. The protostomes are divided into the **lophotrochozoans**, including annelids, molluscs, and brachiopods, many of which share a trochophore larval stage in their life cycle; and the **ecdysozoans**, or molting animals, including the arthropods, onychophora, nematodes, and priapulids.

Within these great clades, the branching order has been less well resolved, such that it is unclear which phyla are more closely related. It is worth noting that the recent assignment of arthropods and annelids to two different protostome clades and the grouping of nematodes and arthropods in the same clade are major changes from previous portraits of the animal tree.

The anatomical and developmental features of the Bilateria are very distinct from those of the basal metazoans (cnidara, ctenophores, and porifera). The evolutionary links between basal metazoans and the bilaterians are difficult to conceive. Indeed, as we will see in Chapter 4, major differences exist between the genetic toolkit of these two groups, and the differences are much more substantial than those between most bilaterians.

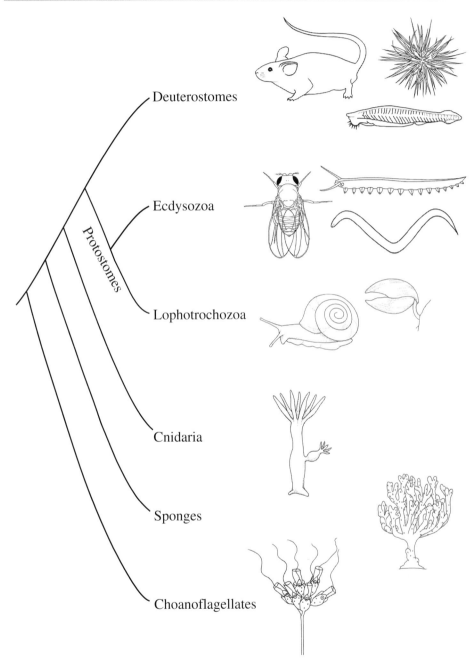

Figure 1.4
Metazoan phylogeny

The current picture of metazoan phylogeny showing representatives of three major bilaterians clades—the deuterostomes, the Lophotrochozoa, and the Ecdysozoa.

Because of the long divergence times between these groups, the phylogenetic relationships between cnidarians, sponges, and ctenophores and the last common ancestor of the Bilateria are uncertain. Many extinct animal lineages, as yet unknown from the fossil record, may have branched off of the metazoan tree between the last common ancestor of all animals and of the Bilateria (see Fig. 1.4).

The gaps in the fossil record; the great differences in anatomy, development, and genome content between radially symmetric animals and bilaterians; and the cryptic early history of bilaterians make inferences about the morphological transformations involved in the origin of animal body plans very speculative. Paleontologists have introduced the concept of **disparity** to refer to differences among body plans and use the term **diversity** to refer to the number of species within a group. The genetic and developmental bases of the morphological diversification of a *particular* body plan within a phylum are far more accessible than are the origins of *different* body plans. Therefore, we will focus on evolutionary trends within a few select phyla, such as the arthropods and chordates, making the implicit assumption that the same sort of genetic mechanisms involved in the evolution of large-scale morphological diversity within phyla also gave rise to fundamental differences in body plans.

GENERAL FEATURES OF ANIMAL DESIGN AND DIVERSITY

One of the most outstanding features of animal design, particularly of larger bilaterians, is their construction from repeating structures (or modules). The segments of arthropods and annelids and the vertebrae (and associated processes) of vertebrates are the basic units of body plan organization in these phyla (Fig. 1.5a–c). Similarly, many body parts such as the insect wing (Fig. 1.5d) and the **tetrapod** hand (Fig. 1.5e) are composed of repeated structures.

An important trend in the morphological evolution of animals has been the **individualization** of modular elements. For example, among the arthropods, we observe a large number of different segment types in crustaceans and insects. This diversity far exceeds that found in the **onychophora**, a phylum closely related to the arthropods. Thus the evolution of the onychophoran/arthropod clade has been marked by increased diversity of segment types from the more uniform patterns found in earlier forms. Similarly, in some mammals, teeth are differentiated into molars, premolars, canines, and incisors, whereas in the ancestral condition exhibited by most reptiles, the teeth are of uniform shape. Because the divergence in the number, morphology, and function of these repeated units characterizes many of the large-scale differences that distinguish related taxa, understanding how repeated structures form and become individualized is a prerequisite for understanding the developmental basis of large-scale morphological evolution.

The modular organization of animal bodies and body parts has long been recognized by comparative biologists. William Bateson, in his classic treatise *Materials for the Study of Variation* (1894), identified several kinds of organization found among animals. More importantly, he was the first to bring a Darwinian perspective to the question of how different body patterns may have evolved. Bateson focused particularly on the repetition of parts, cataloguing a large number of rare, but naturally occurring variants that differed from the norms within various species with regard to either the number or individualization of characters. He suggested that these variations within species could provide insight into the evolution of the large-scale morphological discontinuities between species. For example, variations in the number of body segments within onychophora and centipede species, and of vertebrae in humans and

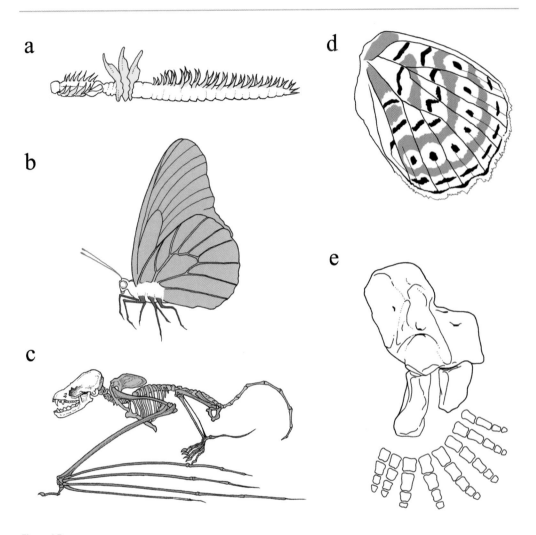

Figure 1.5
The modularity of body plans and body parts

The body plans of many major phyla, including the annelids (**a**), arthropods (**b**), and chordates (**c**) are composed of many repeating parts. Some of these parts are similar or identical in appearance to other parts; others are individuated. Sets of serially homologous structures are shaded a unique color. Body parts, such as a butterfly wing (**d**), or a fossil tetrapod limb from the amphibian fossil *Acanthostega* (**e**) are also composed of repeating structures or patterns, some of which are differentiated from others. For example, *Acanthostega* has eight digits, but like its modern descendants, only five digit types are differentiated.

Source: Parts a–c from Weatherbee SD, Carroll SB. Cell 1999;97:283–386; part e from Michael Coates.

pythons, suggested to Bateson that such discontinuities arose at some frequency in populations and therefore represented plausible steps in the morphological diversification of species.

The question of whether evolution may progress in large, discrete steps remains controversial (we will address this issue in Chapter 7). Nevertheless, these sorts of variants and the organizational concepts espoused by Bateson have been enormously helpful in understanding the genetics and developmental logic underlying the modularity of animal design. In fact, they led to the discovery of genes that play key roles in morphological evolution, albeit not in the fashion Bateson first imagined.

Four fundamental kinds of large-scale, evolutionary changes in morphology are most prevalent in modularly organized animals and are the most significant in terms of adaptation:

1. *Changes in the number of repeated parts* Bateson referred to this type of change as **meristic** variation when describing differences within species. Differences in segment number and vertebral number are some of the most obvious characteristics that distinguish classes of arthropods and various classes and orders of vertebrates, respectively (Fig. 1.6).

2. *Diversification of serially homologous parts* A series of reiterated parts are termed **serially homologous**. The individualization of repeated parts in a lineage results in the differentiation of serially homologous structures. For example, arthropod appendages are serially homologous structures. In the course of arthropod evolution, ancestrally similar appendages have evolved into antennae, various mouthparts, walking legs, and genital structures. In vertebrates, serially homologous vertebrae have evolved into distinct cervical, thoracic, lumbar, and sacral vertebral types.

3. *The diversification of homologous parts* One of the most prevalent trends in animal evolution is the morphological diversification of **homologous** parts between lineages. The same structures in different lineages are termed "homologous" when they share a common history, even if they no longer serve the same function. For example, all tetrapod forelimbs are homologous (Fig. 1.7). Despite their differing appearances and functions, bird wings, bat wings, and human forelimbs have all conserved the basic architecture of the tetrapod forelimb.

4. *The evolution of novelties* New characters or "novelties" may arise from a preexisting structure or evolve de novo and become adapted to a new purpose. The evolution of feathers, fur, teeth, antlers, and butterfly wing eyespots are examples of such morphological novelties.

Considering that modularly organized animals are among the most diverse groups (in terms of both the number and morphology of species), could there be a correlation between body design and evolutionary diversity? One possible explanation for this relationship is that modular organization allows one part of the animal to change without necessarily affecting other parts. The evolution of genetic mechanisms that control the individualization of parts would allow for the uncoupling of developmental processes in one part of the body from the developmental processes in another part of the body. In this fashion, for example, vertebrate forelimbs can evolve into wings while hindlimbs remain walking legs. Dissociation of the forelimb and hindlimb developmental programs allows further modifications to occur selectively in either structure, such as the development of feathers in the forelimb of birds and scales in the hindlimb.

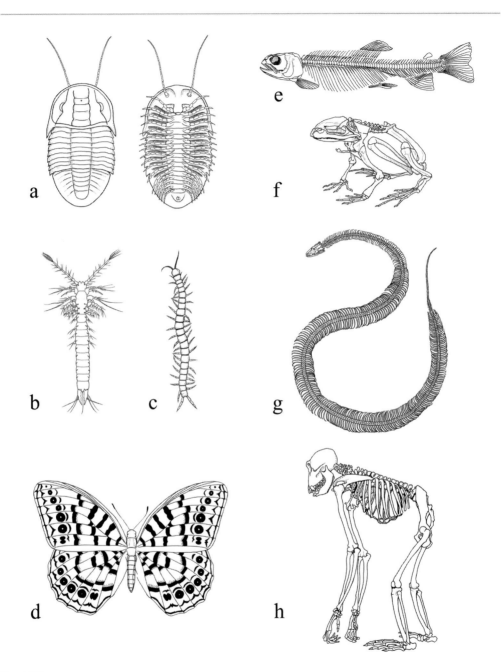

Figure 1.6
Meristic differences among arthropods and among vertebrates

Among arthropods such as this trilobite (**a**), crustacean (**b**), centipede (**c**), and insect (**d**), the number of body segments differs, as does the diversity of segment morphology. Among vertebrates, the number of vertebrae and associated processes differs considerably between a fish (**e**), frog (**f**), python (**g**), and chimpanzee (**h**).

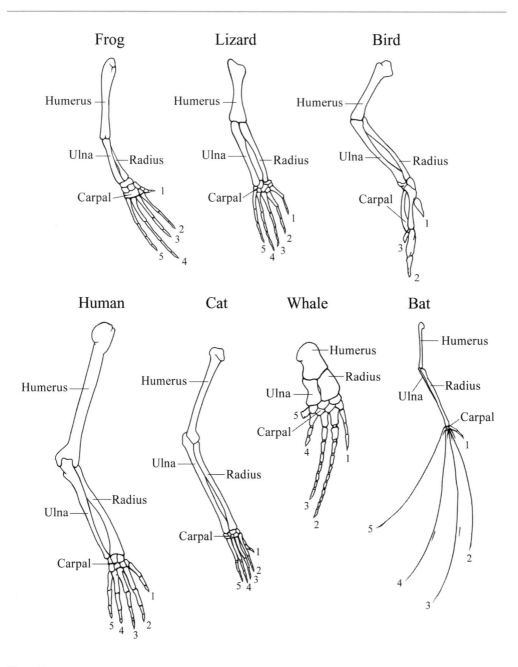

Figure 1.7

The diversification of homologous parts

All vertebrate forelimbs are homologous structures whose anatomy has undergone considerable diversification in the evolution and adaptation of these various vertebrate lineages.

Source: Redrawn from Ridley M. Evolution, 2nd ed. Malden, MA: Blackwell Science, 1996.

EVOLUTION AND DEVELOPMENT: DNA AND DIVERSITY

To understand the major trends in animal diversity and the various kinds of morphological evolution, we must first understand how animal form is generated. Morphology is the product of development, the process through which a single fertilized egg cell gives rise to an entire organism. The physical basis of animal diversity has been viewed since Darwin's time as the outcome of development. Until very recently, however, the developmental principles underlying animal design remained unknown. Although experimental embryologists of the late 1800s and the first half of the 1900s had identified many fascinating phenomena concerning the organization of embryos and the formation of particular structures, the mechanisms responsible for these properties were beyond their reach.

With better understanding of the nature of genes and the process of gene regulation, development has been increasingly viewed as a process orchestrated by the products of genes. Thus the puzzles of embryology, such as how cells come to know their position and identity within a developing animal, have become rephrased in genetic terms. Given that the DNA of (most) all cells in an animal is identical, how do different cells acquire the unique morphologies and functional properties required in the diverse organs and tissues of the body? We now understand that this process occurs through the selective expression of distinct subsets of the many thousands of genes in any animal's genome in different cells. How genes are turned on and off in different cells over the course of animal development is an exquisitely orchestrated regulatory program whose features are only now coming into detailed view.

If morphological diversity is all about development, and development results from genetic regulatory programs, then is the evolution of diversity directly related to the evolution of genetic regulatory programs? Simply put, yes. But to understand how diversity evolves, we must first understand the genetic regulatory mechanisms that operate in development. In other words, what is the genetic toolkit of development and how does it operate to build animals? In the next two chapters, we will examine some of the general features of the genetic and regulatory logic of animal development.

SELECTED READINGS

The Early Fossil Record

Conway, Morris S. The fossil record and the early evolution of the Metazoa. *Nature* 1993; 361:219–225.

———. The crucible of creation: the Burgess shale and the rise of animals. Oxford, UK: Oxford University Press, 1998.

Fortey, R. The Cambrian evolutionary "explosion": decoupling cladogenesis from morphological disparity. *Biol J Linnean Soc* 1996; 5713–5733.

Glaessner, M. F. The dawn of animal life: a biohistorical study. Cambridge, UK: Cambridge University Press, 1984.

Gould, S. J. Wonderful life: the Burgess shale and the nature of history. New York: Norton, 1989.

Knoll, A. H., Carroll, S. B. Early animal evolution: emerging views from comparative biology and geology. *Science* 1999; 284:2129–2137.

Valentine, J. W., Jablonski, D., Erwin, D. H. Fossils, molecules and embryos: new perspectives on the Cambrian explosion. *Development* 1999; 126:851–859.

Molecular Clocks and Animal Evolution

Ayala, F. J., Rzhetsky, A. Origin of the metazoan phyla: molecular clocks confirm paleontological estimates. *Proc Natl Acad Sci USA* 1998; 95:606–611.

Wray, G., Levinton, J., Shapiro, L. Molecular evidence for deep precambrian divergences among metazoan phyla. *Science* 1996; 274:568–573.

Animal Phylogeny

Adoutte, A., Balavoine, G., Lartillot, N., de Rosa, R. Animal evolution: the end of the intermediate taxa? *Trends in Genetics* 1999; 15:104–108.

Aguinaldo, A. M. Evidence for a clade of nematodes, arthropods, and other moulting animals. *Nature* 1997; 387:489–493.

Maley, L. E., Marshall, C. R. The coming of age of molecular systematics. *Science* 1998; 279:505–506.

Animal Design and Evolution: Modularity and Homology

Bateson, W. Materials for the study of variation. London: Macmillan, 1894.

Gilbert, S., Opitz, J., Raff, R. Resynthesizing evolutionary and developmental biology. *Develop Biol* 1996; 173:357–372.

Raff, R. The shape of life. Chicago: University of Chicago Press, 1996.

Wagner, G. P. Homologues, natural kinds and the evolution of modularity. *Am Zoologist* 1996; 36:36–43.

———. The origin of morphological characters and the biological basis of homology. *Evolution* 1989; 43:1157–1171.

The Genetic Toolkit for Development

The foremost challenge for embryology has been to identify the genes and proteins that control the development of animals from an egg into an adult. Early embryologists discovered that localized regions of embryos and tissues possess properties that have long-range effects on the formation and patterning of the primary body axes and appendages. Based on these discoveries, they postulated the existence of substances responsible for these activities. However, the search for such molecules proved fruitless until the relatively recent advent of genetic and molecular biological technologies. The most successful approach to understanding normal development has involved the isolation of single gene mutants that have discrete and often large-scale effects on body pattern.

In this chapter, we take an inventory of the essential genetic toolkit for animal development. We concentrate on genes first discovered in insects, where systematic screens for developmental genes were pioneered. Importantly, however, it turns out that these genes are present in many other animals. We describe how members of the genetic toolkit were identified and what kinds of gene products they encode. In addition, we illustrate the general correlation between these genes' patterns of expression with the development of the morphological features they affect. Finally, we briefly survey their distribution and function in other animals.

Only a small fraction of all genes in any given animal constitute the toolkit that is devoted to the formation and patterning of the body plan and body parts. Two classes of gene products with the most global effects on development are of special interest: families of proteins called transcription factors that regulate the expression of many other genes during development, and members of signaling pathways that mediate short- and long-range interactions between cells. The expression of specific transcription factors and signaling proteins marks the location of many classically defined regions within the embryo. These proteins control the formation, identity, and patterning of most major features of animal design and diversity.

BEFORE THE TOOLKIT—ORGANIZERS, FIELDS, AND MORPHOGENS

Long before any genes or proteins affecting animal development were characterized, embryologists sought to identify the basic

"The only way in which we may hope to get at the truth is by the organization of systematic experiments in breeding, a class of research that calls perhaps for more patience and more resources than any other form of biological inquiry. Sooner or later such investigation will be undertaken and then we shall begin to know."

— W. Bateson,
Material for the Study
of Variation (1894)

" . . . if the mystery that surrounds embryology is ever to come within our comprehension, we must . . . have recourse to other means than description of the passing show."

—T. H. Morgan,
Experimental
Embryology (1927)

principles governing animal design. In their search, they focused on the large-scale organization of the primary body axes, the differentiation of various **germ layers** (ectoderm, mesoderm, and endoderm), and the polarity of structures such as appendages and insect segments. By manipulating embryos and embryonic tissues, primarily by transplantation and ablation, researchers discovered many important properties of developing embryos and tissues. Much of the fascination of embryology stems from the remarkable activities of discrete regions within developing embryos in organizing the formation of body axes and body parts. Furthermore, these classical concepts of embryonic organization present a very useful framework for considering how that organization can change during evolution. We will briefly review some of these experiments and ideas before addressing their genetic and molecular manifestations.

The first demonstration of **organizers**—regions of embryos or tissues that have long-range effects on the fate of surrounding tissues—was achieved by Mangold and Spemann in 1924. They transplanted the lip of the blastopore, the invagination where mesoderm and endoderm move inside the amphibian embryo, of a newt gastrula into another newt embryo and found that the transplanted tissue could induce a second complete body axis (Fig. 2.1a). The additional embryo induced was partly derived from the transplanted graft and partly derived from the host. The equivalent of the "Spemann organizer" in amphibians has been found in chick and mouse embryos, and it is now recognized to be a structure characteristic of all chordate embryos.

Other organizers with long-range effects on surrounding tissues have been identified in the developing vertebrate limb bud. Transplantation of a discrete patch of posterior tissue to an ectopic anterior site induces the formation of limb structures (digits, tendons, muscles) with mirror-image polarity to the normal anteroposterior order (Fig. 2.1b). In contrast, transplantation or removal of anterior tissue has no effect on limb development, suggesting that this posterior region of the limb bud, dubbed the **zone of polarizing activity (ZPA)**, organizes anteroposterior (that is, the thumb-to-pinkie axis) polarity and limb formation.

Another organizer operates from the most distal tip of the limb bud, the **apical ectodermal ridge (AER).** Removal of this region truncates the limb and deletes distal elements (digits), whereas transplantation of the AER to an early limb bud can induce outgrowth of a limb (see Fig. 2.1b).

One explanation for the long-range polarizing and inductive effects of the Spemann organizer, ZPA, and AER is that these tissues are sources of inducer molecules, or **morphogens**—that is, substances whose concentrations vary and to which surrounding cells and tissues respond in a concentration-dependent manner. The response to a morphogen depends, then, on the distance of the responding tissue from the source. For example, if the ZPA is a source of a morphogen, then diffusion of this substance can establish a **gradient** of inducer concentration. Induction of different digit types depends on the morphogen concentration, with low levels of morphogen inducing anterior digits (thumb) and high levels inducing posterior digits (pinkie) (see Fig. 2.1b).

Organizers have been demonstrated and morphogens postulated in insects as well as vertebrates. Ligature and cytoplasmic transplantation experiments first suggested that the anteroposterior axis of certain insect embryos is influenced by two organizing centers, one at each pole of the egg (Fig. 2.1c), that behave as sources of morphogens. Similarly, the polarity of cells within insect segments appears to be organized by signals that produce a graded pattern (Fig. 2.1d).

One difficulty with this picture of morphogen-producing organizers arises when we attempt to explain the boundaries of their range of influence. All of the cells in a growing

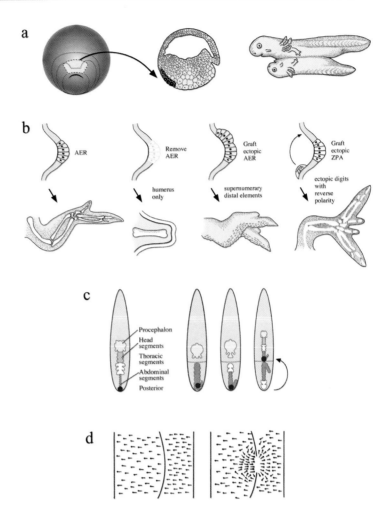

Figure 2.1
Organizers in animal embryos

Transplantation and ablation experiments have been used to investigate the long-range organizing activities of embryonic tissues. (**a**) The Spemann organizer. The dorsal blastopore lip of an early amphibian embryo can induce a second embryonic axis and embryo when transplanted to the ventral region of a recipient embryo. (**b**) Limb organizers. The apical ectodermal ridge (AER) is required for formation of distal limb elements. Removal leads to loss of structures; transplantation to specific ectopic sites induces extra elements. The zone of polarizing activity (ZPA) organizes the anteroposterior pattern; transplantation to an ectopic site induces extra digits with reverse polarity. (**c**) Insect egg organizer. Ligation of the insect *Euscelis* embryo (marked by the black line) early in development deletes the thorax and abdomen; later ligations leave more segments intact. However, transplantation of the posterior pole cytoplasm (marked by the black dot) into the anterior of a ligated embryo induces the formation of a complete embryo. This result demonstrates that the posterior cytoplasm has organizer activity. (**d**) Within insect segments, epithelial polarity is organized by signaling sources. Ablation of a segment boundary (indicated by the interruption of the black line) reorganizes segment polarity (indicated by the orientation of small black hairs).

Source: Parts a–c redrawn from Gilbert S. Developmental biology, 5th ed. Sunderland: Sinauer Associates, 1997; part d from Lawrence PA. The making of a fly. Oxford, UK: Blackwell Science, 1992.

embryo are in contact with other cells, so how is it that some parts respond and others do not? One explanation involves the concept of the **morphogenetic field**. Early embryologists demonstrated that some parts of developing animals, such as the forelimb field, could be transplanted to another site and still differentiate properly—that is, into a forelimb. In addition, if undetermined cells were introduced into the field, they could become incorporated into the limb. These transplantable, self-regulating fields are discrete physical units or modules of embryonic development. They form bounded domains within which specific programs of morphogenesis occur. The term "primary field" applies to the entire embryo before the axes are determined; the limbs, eyes, and other organs are termed "secondary fields," or organ **primordia**.

Secondary fields may be further subdivided into "tertiary fields," defined by physical or developmental boundaries. **Compartments** are one special type of subdivision. First demonstrated within the wing imaginal disk of the fruit fly *Drosophila melanogaster,* compartments are composed of populations of cells that do not intermix with cells outside the compartment.

Further progress in understanding the nature of organizers, morphogens, and fields stalled after their discovery and description in the first half of the 1900s. The impasse was ultimately broken by the discovery of genes whose products governed the activity of organizers, behaved as morphogens, and controlled the formation and identity of embryonic fields. These genes make up the "toolkit" for animal development.

THE GENETIC TOOLKIT

Animal genomes contain thousands of genes. Many of these genes encode proteins that function in essential processes in all cells in the body (for example, metabolism, biosynthesis of macromolecules) and are often referred to as "housekeeping genes." Other genes encode proteins that carry out specialized functions in particular cells or tissues within the body (for example, oxygen transport, immune defense) or, to extend the housekeeping metaphor, in specific "rooms" in the "house." But here we are interested in a different set of genes, those whose products govern the construction of the house—the toolkit that determines the overall body plan and the number, identity, and pattern of body parts.

Toolkit genes have generally first been identified based on the catastrophes or monstrosities that arise when they are mutated. Two sources of toolkit gene mutations exist. The first source comprises rare, spontaneous mutations that arise in laboratory populations of model animals (for example, fruit flies, mice). The second source consists of mutations induced at random by treatment with mutagens (such as chemicals or radiation) that greatly increase the frequency of damaged genes throughout the genome. Elegant refinements of the latter approach, particularly in the fruit fly *Drosophila melanogaster,* have enabled systematic searches for members of the genetic toolkit for animal development.

Intensive screens for genes that affect the formation of the insect embryonic and adult body patterns and analysis of the structure, function, and expression of the proteins they encode have revealed several critical features of the genetic toolkit for development:

1. *The toolkit is composed of a small fraction of all genes*   Only a small subset of the entire complement of genes in the genome affects development in discrete ways.

2. *Most toolkit genes encode either transcription factors or components of signaling pathways*   Therefore, toolkit genes generally act, directly or indirectly, to control the expression of other genes.

3. *The spatial and temporal expression of toolkit genes is often closely correlated with the regions of the animal in which the genes function.*

4. *Toolkit genes can be classified according to the phenotypes caused by their mutation.* Similar mutant phenotypes often reflect genes that function in a single developmental pathway. Distinct pathways exist for the generation of body axes, for example, and for the formation and identity of fields.

5. *Many toolkit genes are widely conserved among different animal phyla.*

Because the discovery of the insect toolkit has offered a direct path to identifying developmental genes in other animals, we will begin our inventory of the genetic toolkit for animal development by considering *Drosophila melanogaster.*

The Drosophila *toolkit*

Classifying genes according to their developmental function. Many mutations have been isolated that alter the embryonic and/or adult body pattern of *Drosophila*. It has proved very useful to group the genes affected by these mutations into several categories based on the nature of mutant phenotypes. Most toolkit genes can be classified according to their function in controlling the identity of fields (for example, different segments and appendages), the formation of fields (for example, organs and appendages), the formation of cell types (for example, muscle and neural cells), and the specification of the primary body axes.

We begin by considering the genes that control the identity of segments and appendages. This choice is made partly for historical reasons and partly to follow a hierarchical approach. The genes controlling field identity were among the very first toolkit members discovered, and their identification inspired much of the genetic and molecular biological innovations that catalyzed the discovery of the rest of the toolkit. In addition, they are among the most globally acting developmental genes that affect animal form. Next, we discuss genes that control the patterning of fields at progressively finer scales, from the formation of entire fields, to compartments within fields, and then to differentiated cell types.

Homeotic genes and segmental identity. Among the most fascinating kinds of abnormalities to be described in animals are those in which one normal body part is replaced with another. Bateson catalogued several oddities of this nature, coining the term **homeotic** to describe such transformations. Among the most common homeotic variants noted by Bateson were arthropods in which one type of appendage formed in the position of another, such as a leg in place of an antenna (Fig. 2.2a), and vertebrates in which one type of vertebra or rib replaced another, such as a thoracic vertebra in place of a cervical vertebra (Fig. 2.2b).

Intriguing as Bateson's specimens were, most were one-of-a kind museum pieces in which only one member of a bilateral pair of structures was affected. To carry out a thorough investigation of the phenomenon of homeosis and its genetics, researchers required mutants that would breed true in subsequent generations. In 1915, Calvin Bridges isolated a spontaneous mutation in *Drosophila* in which part of the haltere (the posterior flight appendage in flies) was transformed into wing tissue. In this mutant, dubbed *bithorax,* part of a structure that forms on the third thoracic segment develops as the corresponding structure on the second thoracic segment. The haltere and wing are serially homologous

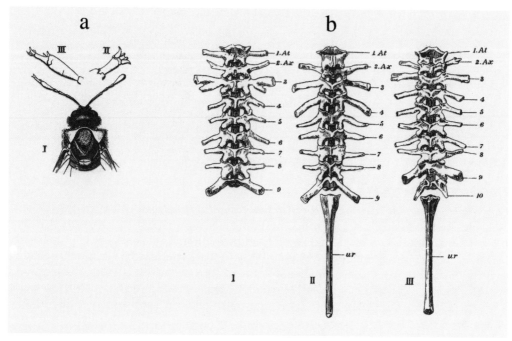

Figure 2.2
Homeotic transformations in an arthropod and a vertebrate

(a) Homeosis in the insect *Cimbex axillaris*, with the left antenna being transformed toward leg identity. (b) Homeosis in a frog. The middle specimen is normal. The specimen on the left has processes emanating from the atlas (top of vertebral column). The specimen on the right has an extra set of vertebrae.

Source: Bateson W. Materials for the study of variation. London: Macmillan, 1894.

appendages, so the *bithorax* mutation causes the partial transformation of the identity of one structure (the haltere) into its serial homolog (the wing). A more complete transformation of the entire haltere into a wing can occur if additional mutations are combined with *bithorax,* producing a four-winged fly (Fig. 2.3a,b).

In the following decades, several more homeotic mutants were identified in *Drosophila*, and in other insects as well. All of these homeotic mutations transform the identities of segments and their associated structures into those of other segments. For example, certain *Antennapedia* mutations cause the transformation of antennae into legs (Fig. 2.3c), which are also serial homologs. The direction of the homeotic transformations depends on whether a mutation causes a loss of homeotic gene function where the gene normally acts, or a gain of homeotic gene function in places where the homeotic gene does not normally act. For example, *Ultrabithorax* (*Ubx*) acts in the haltere to promote haltere development and repress wing development. Loss-of-function mutations in *Ubx* transform the haltere into a wing. Dominant mutations that cause *Ubx* to gain function in the wing transform that structure into a haltere. Similarly, the antenna-to-leg transformations of *Antennapedia* mutants reflect a dominant gain of *Antennapedia* gene function in the antenna.

Figure 2.3
Homeotic mutants of *Drosophila melanogaster*

(**top**) Normal fly with one pair of wings on T2 and halteres on T3. (**middle**) Triple mutant for three mutations in the *Ultrabithorax* gene abolishes *Ubx* function in the posterior thorax and causes the appearance of an extra set of wings (transformation of T3 → T2 identity). (**bottom**) *Antennapedia* mutant in which the antennae are transformed into legs.

The fascination with homeotic mutants stems from two issues. First, it is startling that a single gene mutation could change entire developmental pathways so dramatically in a complex animal. Second, it is curious that the structure formed in the mutant is a well-developed likeness of another body part.

More detailed understanding of homeotic gene function was made possible by some particularly ingenious methods for analyzing the effects of mutations on the behavior of a group of cells in otherwise normal (or "wild-type") tissues. That is, rather than being limited to examining the effect of homeotic mutations on whole animals, the behavior of clones of mutant cells could be observed within otherwise normal animals (Fig. 2.4). This technique was used to determine that the effects of homeotic mutations generally remain limited to cells with mutant genotypes; such behavior is termed cell **autonomous**. Thus a patch of cells in the haltere that lacks *Ubx* function forms wing tissue, even when it is surrounded by normal haltere cells (see Fig. 2.4). This finding suggested that homeotic genes act within cells to select their developmental fate. Homeotic genes, and other genes with analogous functions in controlling cell fate, are therefore known as **selector** genes.

Although homeotic genes were first identified through spontaneous mutations affecting adult flies, they are required throughout most of *Drosophila* development to determine segmental identity. Systematic screening for homeotic genes led to the identification of

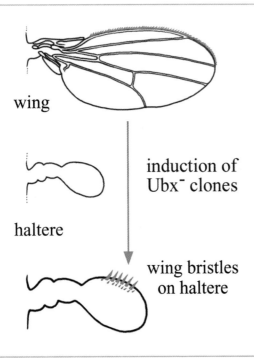

wing

induction of
Ubx⁻ clones

haltere

wing bristles
on haltere

Figure 2.4
Cell autonomy of homeotic mutations

The *Drosophila* wing and haltere have different pattern elements, such as the occurrence of sensory bristles at the leading edge of the wing (red). Clones lacking *Ubx* function in the haltere form wing structures (for example, the sensory bristles shown in red) in positions corresponding to those of the wing.

Source: Redrawn from Lawrence PA. The making of a fly. Oxford, UK: Blackwell Scientific, 1992.

eight linked genes, collectively referred to as **Hox** genes, that affect the specification of particular segment identities in the developing *Drosophila* embryo, larva, and adult. In addition to *Ultrabithorax* (*Ubx*) and *Antennapedia* (*Antp*), they include *labial* (*lab*), *proboscipedia* (*pb*), *Deformed* (*Dfd*), *Sex combs reduced* (*Scr*), *abdominal-A* (*abd-A*), and *Abdominal-B* (*Abd-B*). Generally, the complete loss of any *Hox* gene function causes transformations of segmental identity and is lethal in early development. The spontaneous homeotic mutants found in viable adults are caused by partial loss of gene function or are dominant such that in heterozygotes normal gene function is provided by the wild-type allele.

One of the most intriguing features of these *Hox* genes is that they are linked in two **gene complexes** in *Drosophila*, the Bithorax and Antennapedia Complexes; each complex contains several distinct homeotic genes. Furthermore, the order of the genes on the chromosome and within the two complexes corresponds to the rostral (head) to caudal (rear) order of the segments that they influence, a relationship described as **colinearity** (Fig. 2.5).

The relationship between the structure of *Hox* gene complexes and the phenotypes of *Hox* mutants was illuminated by the molecular characterization of both the Bithorax Complex and the Antennapedia Complex. Cloning of the *Hox* genes provided the means to uncover when and where each of the eight genes is expressed during development. The ability to visualize *Hox* and other gene expression patterns during development was crucial to understanding the correlation between gene function and phenotypes. Localization of *Hox* genes' RNA transcripts by in situ hybridization or of *Hox* proteins via immunological methods (Fig. 2.6) revealed that all *Hox* genes are expressed in spatially restricted,

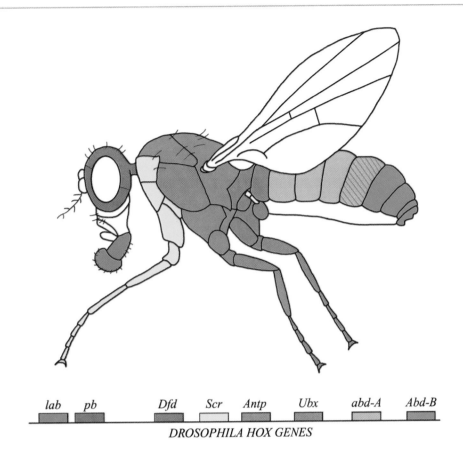

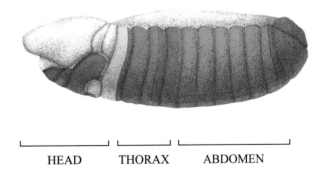

Figure 2.5
The *Hox* genes of *Drosophila*

Eight *Hox* genes regulate the identity of regions within the adult (**top**) and embryo (**bottom**). The color coding represents the segments and structures that are affected by mutations in the various *Hox* genes.

Source: Modified from Carroll SB. Nature 1995;376:479–485.

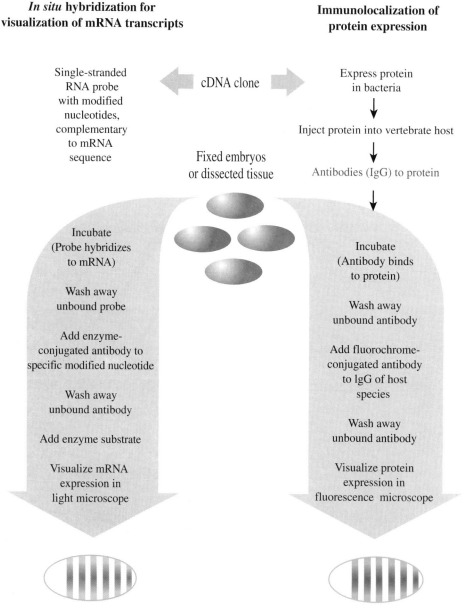

Figure 2.6
Methods for visualizing gene expression in developing animals

The two most common means of visualizing where a gene is transcribed and its protein product is synthesized are (**left**) in situ hybridization of complementary RNA probe to mRNA and (**right**) immunolocalization of protein expression. The procedures for each method are indicated. Gene expression patterns are visualized as the product of enzymatic reactions (**left**) or with fluorescently labeled compounds (**right**).

sometimes overlapping domains within the embryo. These genes are also expressed in subsets of the developing larval **imaginal discs**, which proliferate during larval development and differentiate during the pupal stages to give rise to the adult fly.

The patterns of *Hox* gene expression generally correlate with the regions of the animal affected by homeotic mutations. For example, the *Ubx* gene is expressed within the posterior thoracic and most anterior abdominal segments of the embryo (Fig. 2.7a). The development of these segments is altered in *Ubx* mutants. In larvae, *Ubx* is expressed in the developing haltere, but not in the developing wing (Fig. 2.7b–e). This expression correlates with the requirement for *Ubx* to promote haltere development and to suppress wing identity.

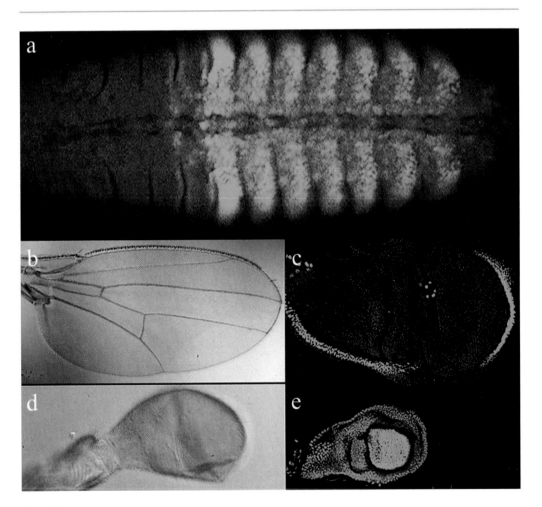

Figure 2.7
Hox **gene expression**

Hox gene expression is restricted to regions of the body and to particular structures. For example, (**a**) the Ubx protein (shown in green) is expressed in the posterior thoracic and seven anterior abdominal segments of the embryo. (**b**) The adult wing. (**c**) Ubx is not expressed in the cells of the wing imaginal disc. (**d**) The adult haltere. (**e**) Ubx is expressed in the cells of the haltere imaginal disc.

The boundaries of *Hox* gene expression in the *Drosophila* embryo are not segmental, but usually begin in the posterior part of one segment and extend to (or beyond) the anterior portion of the next-most-posterior segment, a unit dubbed a **parasegment**. In the various imaginal tissues of the developing adult, homeotic genes are often expressed in segmental domains. For example, flies have three pairs of legs, with one pair extending from each of the three thoracic segments. Each pair of adult legs has a distinctive morphology. Indeed, genetic analysis has shown that the morphology of the first legs is largely influenced by the *Scr* gene, the second legs by the *Antp* gene, and the third legs by the *Ubx* gene. These respective genetic requirements correlate with the respective patterns of homeotic gene expression in the developing imaginal legs.

It is crucial to understand the distinction between *Hox* gene function in determining the *identity* of a field, as opposed to a requirement for *Hox* gene function in the *formation* of the field. The antennae, mouthparts, and walking limbs of flies all develop from serially homologous arthropod limb-type fields. In the absence of homeotic genes, each limb field develops, but with antennal identity. Therefore, *Hox* genes specify the particular identity, but are not required for the formation of the limb fields. The expression and function of *Hox* genes are not limited to body segments and their appendages. These genes act as **region-specific selectors** in all three germ layers (ectoderm, mesoderm, and endoderm) and in very diverse structures and tissues.

The homeobox. The large effects of *Hox* genes on the developmental fates of entire segments and structures made the nature of the proteins encoded by these genes of special interest. The close genetic linkage and similar function of *Drosophila Hox* genes suggested that they might have evolved through the tandem duplication of one or more ancestral *Hox* genes. This idea led to the discovery that *Hox* genes of the Bithorax (BX-C) and Antennapedia (ANT-C) Complexes were similar enough to hybridize to each other at the DNA level. This similarity was traced to a 180 base-pair (bp) stretch of DNA, dubbed the **homeobox**, that encodes a 60 amino acid protein domain (the **homeodomain**); the sequence of the homeodomain is very similar among the homeotic proteins (Fig. 2.8). The structure of the homeodomain resembles the DNA-binding domain of many prokaryotic regulatory proteins, suggesting that homeotic gene products exert their effects by controlling gene expression during development and that the homeodomain binds to DNA in a sequence-specific manner.

The homeobox gene family is large and diverse. In fact, the homeodomain motif is found in approximately 20 other distinct families of homeobox-containing genes, all of which encode DNA-binding proteins.

Field-specific selector genes. Another class of selector genes acts within specific developing fields to regulate the formation and/or the patterning of entire structures. Such genes have been identified in *Drosophila* through spontaneous or induced mutations that selectively abolish or reduce the development of the eye, wing, limbs, and heart. Several of these genes also have the remarkable ability to induce the formation of ectopic organs when they are expressed at different sites in the animal.

Perhaps the best-known *Drosophila* **field-specific selector** gene is the *eyeless (ey)* gene. Flies that lack *ey* function can reach adulthood, but never develop a compound eye (Fig. 2.9b). Molecular characterization of the *ey* gene revealed that it encodes a member of a particular homeobox gene family (Pax-6), suggesting that the Ey protein acts as a DNA-binding transcription factor to regulate the expression of other genes.

```
lab        NNSGRTNFTNKQLTELEKEFHFNRYLTRARRIEIANTLQLNETQVKIWFQNRRMKQKKRV
pb         PRRLRTAYTNTQLLELEKEFHFNKYLCRPRRIEIAASLDLTERQVKVWFQNRRMKHKRQT
Dfd        PKRQRTAYTRHQILELEKEFHYNRYLTRRRRIEIAHTLVLSERQIKIWFQNRRMKWKKDN
Scr        TKRQRTSYTRYQTLELEKEFHFNRYLTRRRRIEIAHALCLTERQIKIWFQNRRMKWKKEH
Antp       RKRGRQTYTRYQTLELEKEFHFNRYLTRRRRIEIAHALCLTERQIKIWFQNRRMKWKKEN
Ubx        RRRGRQTYTRYQTLELEKEFHTNHYLTRRRRIEMAHALCLTERQIKIWFQNRRMKLKKEI
abd-A      RRRGRQTYTRFQTLELEKEFHFNHYLTRRRRIEIAHALCLTERQIKIWFQNRRMKLKKEL
abd-B      VRKKRKPYSKFQTLELEKEFLFNAYVSKQKRWELARNLQLTERQVKIWFQNRRMKNKKNS
```

```
consensus  -RRGRT-YTR-QTLELEKEFHFNRYLTRRRRIEIAHALCLTERQIKIWFQNRRMK-KKE-
                     Helix 1            Helix 2            Helix 3
```

Figure 2.8
Homeodomains of *Drosophila Hox* genes

Each of the eight *Drosophila Hox* genes encodes proteins containing a highly conserved 60 amino acid DNA-binding domain, the homeodomain, composed of three alpha helices. The third helix is most conserved in sequence. Divergent residues are shaded in red; those shared among subsets of proteins are shaded in blue or green.

The *ey* gene is expressed in the developing eye field in the embryo, and in the larval eye imaginal disc, before the units that make up the compound fly eye form (Fig. 2.10b). Most remarkable, however, is the ability of the *ey* gene when expressed elsewhere in the developing fly, such as in the imaginal wing or leg discs, to induce the formation of eye tissue composed of properly organized, pigmented ommatidia (Fig. 2.9c,d). Its ability to reprogram other developing tissues to form eyes suggests that *ey* is a major regulatory gene in the genetic program of eye development.

The *Distal-less* (*Dll*) gene displays similar properties with respect to the formation of *Drosophila* limbs. Named for the effect of its mutations on the formation of the proximodistal axis of the limbs, *Dll* affects the development of all limbs, including the walking legs, mouthparts, antenna, and genitalia. Complete loss of *Dll* function truncates all limbs (that is, the limbs lack distal elements). The *Dll* gene is yet another type of homeobox-containing gene, suggesting that *Dll* also exerts its effects by regulating the expression of other genes. It is expressed in the limb primordia in the embryo, and in the distal portion of all imaginal limb fields (Fig. 2.10c,d). Expressing *Dll* in places where it is not normally active can induce the outgrowth of ectopic limbs.

Development of the flight appendages of *Drosophila* depends on the function of a pair of genes, *vestigial* (*vg*) and *scalloped* (*sd*), whose products act together in a molecular complex. Fruit flies with *vg* and *sd* mutations lack wings and halteres altogether. The Vg and Sd proteins are expressed in the wing and haltere primordia in the embryo and in fields of cells within the imaginal discs that will give rise to the flight appendages (see Fig. 2.10c). As is the case with the other field-specific selector genes, expression of *vg* with *sd* in developing eyes, legs, antenna, or genitalia can induce the formation of wing tissue. The Vg and Sd proteins form a complex that binds to DNA, indicating that their selector function is mediated by regulation of gene expression.

The formation of the *Drosophila* heart depends on still another selector gene, dubbed *tinman* (*tin*). Mutants lacking *tin* function lack a heart. The *tin* gene is expressed in the developing mesoderm and in all cells that will form the cardiac tissue of the fly. It is a

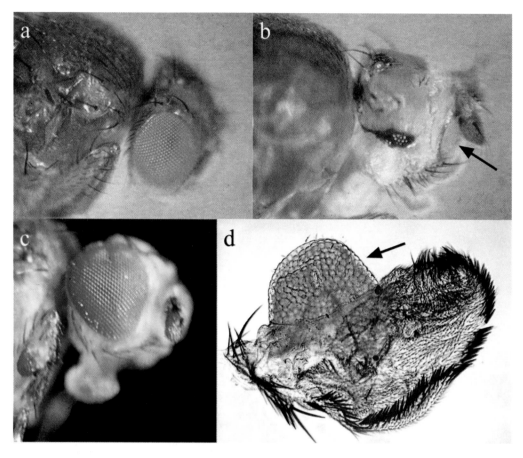

Figure 2.9
The *eyeless* selector gene controls eye development

(a) Normal fly head with eye. (b) The *ey* mutant fly lacks the eye. (c) Expression of the *ey* gene induces the formation of pigmented eye tissue at new sites, including (d) on the wing.

Source: Photographs courtesy of Georg Halder.

member of a distinct homeobox family, and thus also a DNA-binding protein that acts by controlling gene expression.

Compartment selector genes. Several genes have been identified in *Drosophila* that act within certain developing fields to subdivide them into separate cell populations, or compartments. The *engrailed* (*en*) gene acts in the posterior part of all segments of the embryo; it is expressed continuously such that the posterior portions of all structures that develop from these segments also express *en* (Fig. 2.11a). The function of the *engrailed* gene is best understood in the embryo and in the developing wing, where it acts to determine posterior identity. Mutations in this gene cause posterior cells to develop as anterior cells but with reversed segmental polarity, resulting in mirror-image duplications of anterior tissue.

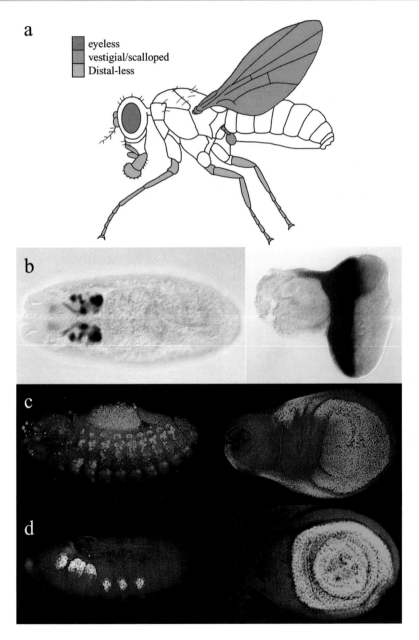

Figure 2.10
Field-specific selector genes

(**a**) Development of parts of the *Drosophila* adult depend upon the function of the *ey* (eyes), *Dll* (limbs), and *vg* (flight appendages) selector genes. (**b–d**) These genes are expressed in both the embryonic primordia (**left**) and larval imaginal discs (**right**), which will give rise to these structures.

Source: Photomicrographs courtesy of Georg Halder and Grace Panganiban.

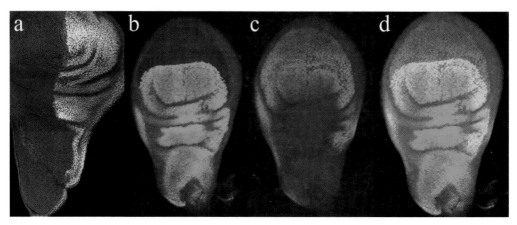

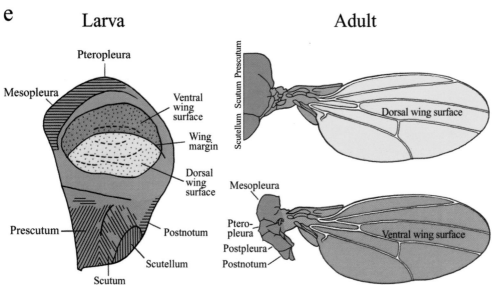

Figure 2.11
Compartmental selector genes
(**a**) The Engrailed protein is expressed in all cells in the posterior compartment of the wing imaginal disc. (**b**) The Apterous protein (shown in green) is expressed in all cells that are in the dorsal compartment of the wing imaginal disc, and subdivides the fields of (**c**) Vestigial-expressing cells into (**d**) dorsal (yellow) and ventral (red) populations. (**e**) The territories marked by expression of the proteins in parts b–d in the larval imaginal disc correspond to future regions of the adult wing.

The *engrailed* gene encodes a distinct class of homeodomain-containing transcription factors.

A second compartmental selector gene, *apterous* (*ap*), subdivides the developing wing imaginal disc into dorsal and ventral compartments (Fig. 2.11b–e). Complete loss of *apterous* function blocks wing development, whereas loss of *apterous* function within a subset of dorsal cells transforms their identity to ventral fate. The Apterous protein belongs to yet another class of homeodomain-containing transcription factors.

Cell-type-specific selector genes. Another class of selector genes operates within developing fields to control the differentiation of particular cell types. The formation of neuroblasts and other neural precursor cells in *Drosophila* requires the action of members of the Achaete-Scute Complex (AS-C), a gene complex that contains four genes. Loss of AS-C gene function in the embryo prevents formation of the nervous system; loss or reduction of individual AS-C gene functions in particular body regions in the imaginal tissues of the developing adult fly causes loss of particular sensory bristles.

All four AS-C genes encode structurally related transcription factors. The genes are expressed in dynamic and complex patterns that foreshadow the formation of central and peripheral nervous system elements in the larva and adult. The development of neural precursors is initiated within clusters of cells that express AS-C genes, from which a single precursor segregates, divides, and gives rise to neurons and associated cells (Fig. 2.12). A similar process involving a distantly related group of transcription factors specifies muscle

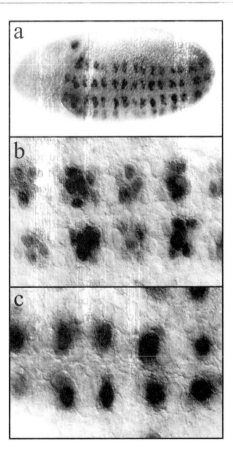

Figure 2.12
A cell-type-specific selector gene

(**a**) The Achaete protein is expressed in clusters of proneural cells (shown in greater detail in part (**b**) that foreshadow the pattern of neural precursors. (**c**) Single precursor cells within each cluster will segregate and give rise to neuroblasts.

development in *Drosophila*. The *twist, nautilus,* and *DMEF-2* genes control the development and differentiation of muscle cells.

Formation of the body axes

***Systematic searches for developmental genes in* Drosophila.** Many of the selector genes described in the previous section were first identified on the basis of the adult phenotypes of spontaneous mutants in *Drosophila*. Most of those mutations, however, did not completely disrupt the gene's function during development. Complete loss of function of many selector genes is lethal at earlier stages of development. Therefore, to find genes that control other aspects of embryo organization and patterning, genetic screens had to be designed that could identify recessive lethal mutations.

Two types of systematic screens have harvested the lion's share of the *Drosophila* genetic toolkit. The first searched for all loci that were required in the fertilized egg, or **zygote**, for proper patterning of the larva. The second type of search sought to identify those genes whose products function in the egg for proper patterning, before the zygotic genome becomes active. Genes whose products are provided by the female to the egg are called maternal effect genes. Mutant phenotypes of strict **maternal effect** genes depend only on the genotype of the female parent (Fig. 2.13).

The two types of systematic mutagenesis screens revealed that mutations in only a small fraction of all genes in the genome have very specific effects on the organization and patterning of the embryo and larva. In addition to their maternal or zygotic actions,

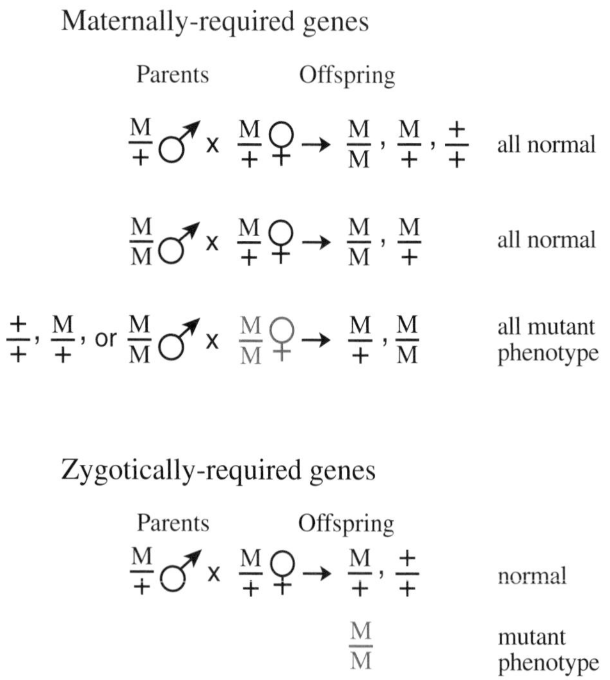

Maternally-required genes

Parents   Offspring

$$\frac{M}{+}\,\male \times \frac{M}{+}\,\female \rightarrow \frac{M}{M}, \frac{M}{+}, \frac{+}{+} \quad \text{all normal}$$

$$\frac{M}{M}\,\male \times \frac{M}{+}\,\female \rightarrow \frac{M}{M}, \frac{M}{+} \quad \text{all normal}$$

$$\frac{+}{+}, \frac{M}{+}, \text{ or } \frac{M}{M}\,\male \times \frac{M}{M}\,\female \rightarrow \frac{M}{+}, \frac{M}{M} \quad \text{all mutant phenotype}$$

Zygotically-required genes

Parents   Offspring

$$\frac{M}{+}\,\male \times \frac{M}{+}\,\female \rightarrow \frac{M}{+}, \frac{+}{+} \quad \text{normal}$$

$$\frac{M}{M} \quad \text{mutant phenotype}$$

Figure 2.13
The genetics of *Drosophila* embryonic development

The phenotypes of offspring depend on either the maternal genotype for maternal effect genes (**top**) or the offspring (zygotic) genotype for zygotically required genes (**bottom**). Mutant (m); wild type (+).

these mutants and the corresponding genetic loci can be classified according to the embryonic axis affected (anteroposterior or dorsoventral), and the type of patterning defect observed. Molecular characterization of these genes identified many of the first known representatives of widely shared transcription factor families and signaling pathways. Indeed, the molecular analysis of many of these genes in *Drosophila* led to the development of tools to isolate them from other animals. Thus the systematic inventory of the *Drosophila* genome is representative of what is generally known about the types of molecules with large-scale effects on animal patterning.

The anteroposterior axis.   A few dozen *Drosophila* genes are required for proper anteroposterior patterning of the embryo and larva. These genes are grouped into five classes based on their realm of influence on embryonic pattern. Each class represents a progressively finer subdivision of the developing embryo.

The first class consists of the maternal effect genes, such as the *bicoid* gene (which affects the anterior region of the embryo) and the *nanos* and *caudal* genes (which affect the posterior region) (Fig. 2.14).

wild type

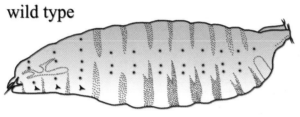

bicoid mutant

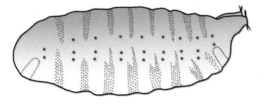

nanos mutant

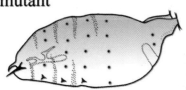

Figure 2.14
Maternal effect genes controlling embryonic polarity in *Drosophila*

(**top**) The cuticle pattern of wild-type *Drosophila* larvae. Various structures develop at characteristic positions along the anteroposterior axis. Note the mouth hooks at the anterior (**left**) end of the animal, the presence of triangular organs in each of the three thoracic segments, and the broad bands of denticles marking each abdominal segment. (**middle**) Embryos from homozygous *bcd/bcd* mutant females lack anterior structures and have duplicated posterior structures. (**bottom**) Embryos from homozygous *nos/nos* mutant females lack most abdominal structures.

Source: Redrawn from Lawrence PA. The making of a fly. Oxford, UK: Blackwell Scientific Publications, 1992.

The second class contains the zygotically active **gap genes**, which include the *hunch-back, Krüppel, giant, knirps, tailless,* and *huckebein* genes, each of which regulates the formation of a contiguous set of segments. Mutations in gap genes lead to gaps in segmentation (Fig. 2.15b).

The third class comprises the **pair-rule genes**, such as *fushi tarazu, even-skipped, hairy,* and *paired*, which act at a double-segment periodicity. Pair-rule mutants display defects in part of each pair of segments (Fig. 2.15c).

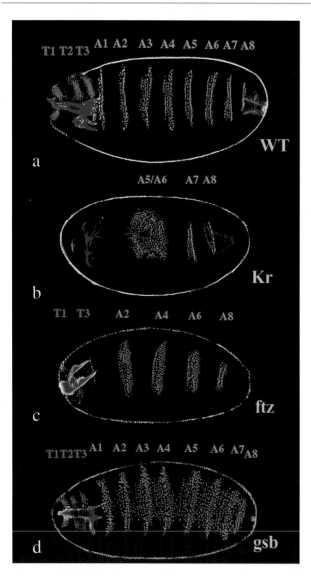

Figure 2.15
Segmentation gene mutants

(**a**) The cuticle of a *Drosophila* larvae has belts or denticles that form characteristic patterns in each segment (T1, T2, T3, A1, and so on). The position and pattern of these denticles are landmarks for identifying developmental abnormalities. (**b**) A *Krüppel* gap mutant. Loss of *Krüppel* function prevents the formation of several segments. (**c**) A *fushi tarazu* pair-rule mutant. Loss of this gene's function results in loss of every other segment boundary and pairwise fusion of segments. (**d**) A *gooseberry* segment polarity mutant. Loss of *gooseberry* function alters the polarity of each segment.

Source: Courtesy of Nipam Patel.

The fourth class contains the **segment polarity genes**, which affect patterning within each segment. Mutants in this class lack the normal polarity of pattern elements within segments and display polarity reversals and segmentation defects (Fig. 2.15d). The products of two segment polarity genes, *wingless* and *hedgehog,* are largely responsible for the polarity-organizing activities identified within insect segments by classical transplantation and ablation techniques.

The fifth class includes the homeotic selector genes, which we discussed earlier in this chapter.

Collectively, these five classes of mutants indicate that the segmental body plan of the *Drosophila* larva is progressively specified by genes acting over the realm of the whole embryo, in subregions of the embryo, in every other segment, and, ultimately, in each individual segment.

The dorsoventral axis. The dorsoventral axis of the *Drosophila* embryo is also regionally subdivided. Several maternal effect genes are required to initiate the establishment of dorsoventral polarity. Embryos from females that lack the activity of any of these genes, such as *dorsal,* are "dorsalized"—that is, ventral structures do not form in these animals. The zygotically active genes *decapentaplegic* (*dpp*), *zerknüllt* (*zen*), *short gastrulation* (*sog*), *twist* (*twi*), and *snail* (*sna*) all play major roles in the subdivision of the dorsoventral axis. In addition, neurogenic genes, such as *Delta* (*Dl*) and *Notch* (*N*), are required in distinct dorsoventral subregions for the formation of the ectoderm.

Expression of toolkit genes. The phenotypes of dead larvae or adult monsters tell only part of the story regarding what developmental genes do. To fully understand the link between these genes and patterning, we must know the timing and location of the genes' expression patterns and the molecular nature of the genes' protein products. The identification of gene classes that, when mutated, cause similar effects in development raises the possibility that a given class of genes may affect the same developmental process or genetic pathway, or that the genes may encode products with similar functions.

Analysis of the expression of toolkit genes has revealed a very informative correlation between the locations at which genes are expressed in development and the pattern of defects caused by mutations. For each of the five classes of anteroposterior axis-patterning genes, a clear correspondence exists between the regions of the embryo in which the gene is transcribed (or the protein product is localized) and the regions affected by mutations in that gene. For example, the *bicoid* and *nanos* proteins are expressed in graded patterns emanating from the anterior and posterior poles of the embryo, respectively (Fig. 2.16a,b). These proteins are largely responsible for the organizing activities that classical experiments identified as residing at the two poles of the insect egg.

The gap genes are expressed in blocks of cells that correspond to the future positions of the segments affected by gap gene mutants (Fig. 2.17a). The pair-rule genes are expressed in one transverse stripe per every two segments, for a total of 7 stripes that span 14 future body segments; the stripes correspond to the periodicity of defects in mutant embryos (Fig. 2.17b). The segment polarity genes are expressed in each segment, in 14 or 15 transverse stripes (Fig. 2.17c). The various dorsoventral patterning genes are expressed in different domains along the dorsoventral axis that correspond to regions of the embryo that give rise to elements such as the mesoderm, neuroectoderm (the part of the ectoderm

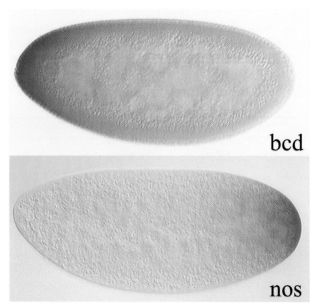

Figure 2.16
Expression of maternal morphogens

The maternally derived Bicoid and Nanos proteins form concentration gradients emanating from the anterior (Bcd) and posterior (Nos) poles of the *Drosophila* embryo.

Source: Photomicrographs courtesy of Ruth Lehmann.

bcd

nos

from which the ventral nervous system develops), and amnioserosa (a dorsal sheet of extraembryonic cells) (Fig. 2.18a–d).

Toolkit gene products: transcription factors and signaling pathway components.
The proteins encoded by selector and axial patterning genes most often belong to one of two categories: transcription factors or components of signaling pathways. Ultimately, these proteins exert their effect through the control of gene expression. Thus developmental processes such as embryonic axis formation and segmentation are organized by regulating gene expression in discrete regions and cell populations of the embryo.

The types of transcription factors found in the *Drosophila* toolkit include representatives of most of the known families of sequence-specific DNA-binding proteins. These families are distinguished by the type of secondary structures in the folded protein that are involved in protein subunit interactions and contact with DNA.

Most transcription factors possess either a **helix-turn-helix**, **zing finger**, **leucine zipper**, or **helix-loop-helix** (HLH) motif (Fig. 2.19a–d; Table 2.1). The homeodomain superfamily belongs to the helix-turn-helix class of factors, for example (see Fig. 2.19a). Three proteins with divergent homeodomain sequences—Bicoid, Fushi tarazu, and Zen—have very different roles in organizing of the anteroposterior axis, segmentation, and dorsoventral axis patterning, respectively. All three proteins are also encoded within the Antennapedia Complex, surrounded by *Hox* genes. Most other homeodomain proteins are encoded by genes dispersed throughout the genome.

Several of the gap genes encode zinc finger proteins, whereas genes such as the dorsoventral patterning gene *twist* and the pair-rule segmentation gene *hairy* encode basic HLH proteins. The identification of these DNA-binding motifs in proteins has allowed biologists to deduce the molecular function of a gene product by inspecting its encoded

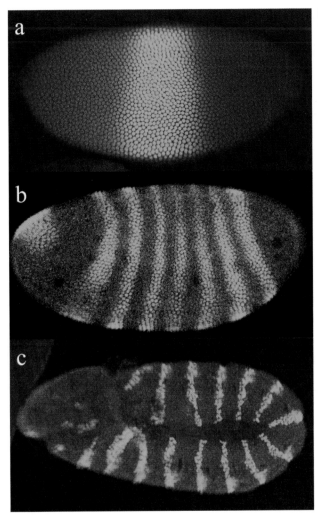

Figure 2.17

Expression of segmentation genes

Drosophila embryos stained with antibodies specific for the (**a**) Krüppel gap protein, (**b**) Hairy pair-rule protein, and (**c**) Engrailed segment polarity protein. Each protein is localized to nuclei in regions of the embryo that are affected by mutations in the respective genes.

Source: Photographs by James Langeland.

sequence, rather than resorting to exhaustive biochemical analysis. Furthermore, because these motifs are involved in contact with DNA, they are often constrained with respect to evolutionary changes in their sequence. As a result, these motifs are useful for isolating gene homologs in other taxa.

The second major category of toolkit genes encode proteins involved in the process of cell signaling, either as ligands, receptors for ligands, or components involved in the intracellular transduction of signals. At least seven major signaling pathways operate in the *Drosophila* embryo: the Hedgehog, Notch, Wingless, Dpp/transforming growth factor-β (TGF-β), Toll, epidermal growth factor (EGF-R) and fibroblast growth factor (FGF-R) signaling pathways (Table 2.2).

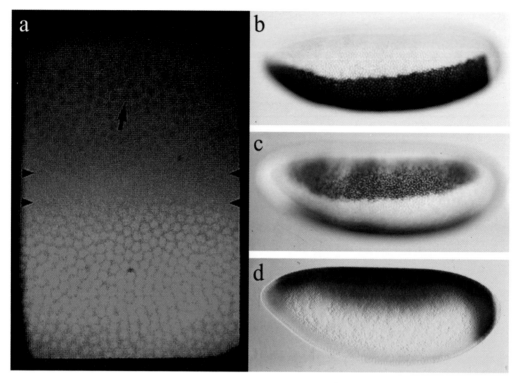

Figure 2.18
Expression of dorsoventral patterning genes

(**a**) Dorsal protein expression. The concentration of Dorsal in cell nuclei is graded from dorsal to ventral cells. Zygotic genes are expressed at different positions along the dorsoventral axis with (**b**) *snail* expression in the most ventral cells, (**c**) *sog* expression in lateral cells, and (**d**) *zen* expression in the most dorsal cells.

Source: Photographs courtesy of Michael Levine.

All of these pathways have a similar construction—namely, each has at least one signaling ligand, at least one receptor spanning the cell membrane, and at least one DNA-binding transcription factor that responds to signaling inputs by binding to target genes to turn them on or off (Fig. 2.20). Although the signaling logic may be similar among pathways, the biochemistry is not. Many structural types of ligands and receptors exist, and each pathway regulates the activity of different types of transcription factors. The response to ligand binding is mediated by a variety of mechanisms that often involve post-translation modifications (for example, protein phosphorylation, proteolysis, binding to a co-factor) that regulate the activity of or the translocation of transcription factors to the cell nucleus (see Fig. 2.20). The similar mutant phenotypes of some developmental genes reflect their involvement in the same signaling pathway. Any loss of a signaling ligand, receptor, or member of the signal transduction machinery will cause the pathway to fail.

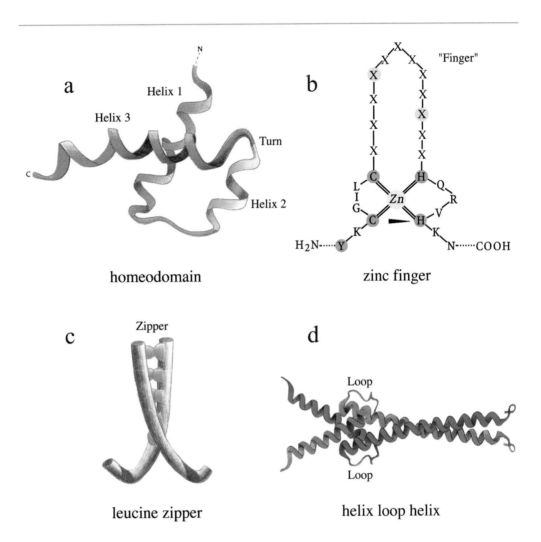

DNA binding motifs

Figure 2.19
Structural motifs of major transcription factor families

(**a**) The second and third α helices of the homeodomain form a helix-turn-helix structure. (**b**) The zing finger motif involves a coordination complex of Zn with critically positioned cysteine (C) and histidine (H) residues. The "finger" contacts DNA. (**c**) The leucine zipper structure is formed by association of two subunits with regularly spaced leucine residues. (**d**) The helix-loop-helix structure is similar to the leucine zipper except that a protein loop interrupts the helices and the association of the subunits.

Pleiotropy of toolkit genes.   Because of the essential roles played by toolkit genes in the initial pattern events of *Drosophila* development, flies with mutations in toolkit genes typically fail to survive the early stages of the life cycle. Most of these genes also play critical roles

TABLE 2.1 *Transcription factors in the* Drosophila *genetic toolkit for development*

Gene/Protein	Developmental Function	Vertebrate Homolog(s)
Homeodomains		
labial	Homeotic	Yes
proboscipedia	Homeotic	Yes
deformed	Homeotic	Yes
Sex combs reduced	Homeotic	Yes
Antennapedia	Homeotic	Yes
Ultrabithorax	Homeotic	No
abdominal-A	Homeotic	No
abdominal-B	Homeotic	Yes
bicoid	Maternal anteroposterior axis organizer	No
caudal	Maternal/zygotic anteroposterior axis organizer	Yes
fushi-tarazu	Pair-rule segmentation	No
even skipped	Pair-rule segmentation	Yes
paired	Pair-rule segmentation	Yes
zerknullt	Dorsoventral axis patterning	Yes
engrailed	Segment polarity/posterior compartment selector	Yes
apterous	Dorsal wing compartment selector	Yes
eyeless	Eye field selector	Yes
Distal-less	Limb field selector	Yes
tinman	Mesoderm, heart selector	Yes
extradenticle	Hox co-factor	Yes
Zinc Finger		
hunchback	Gap segmentation	Not determined
Krüppel	Gap segmentation	Yes
tailless	Gap segmentation	Yes
huckebein	Gap segmentation	Not determined
knirps	Gap segmentation (steroid receptor)	Not determined
odd-skipped	Pair-rule segmentation	Not determined
snail	Dorsoventral axis patterning	Yes
Helix-Loop-Helix		
achaete	Proneural	Yes
scute	Proneural	
lethal of scute	Proneural	
asense	Proneural	
atonal	Proneural (chordotonal)	Yes
nautilus	Myogenic	Yes
Dmef	Myogenic	Yes

TABLE 2.1 *Continued*

Gene/Protein	Developmental Function	Vertebrate Homolog(s)
Helix-Loop-Helix (cont.)		
hairy	Pair-rule segmentation	Yes
twist	Dorsoventral axis patterning	Yes
Other		
dorsal	Dorsoventral axis organizer	Yes
giant	Gap segmentation	Not determined
vestigial	Wing selector	Yes
scalloped	Wing selector	Yes

TABLE 2.2 *Major signaling pathways in the* Drosophila *genetic toolkit*

Pathway	Ligands	Receptors	Transcription Factors	Vertebrate Homologs
TGF-β	decapentaplegic 60A screw	thick veins punt saxophone	Mad Medea	Yes
Wingless	wingless Dwnt 3 Dwnt 5	Dfrizzled-2	DTCF/pangolin	Yes
Notch	Delta Serrate (Fringe)	Notch	Suppressor of Hairless	Yes
Hedgehog	hedgehog	patched smoothened	cubitus interruptus	Yes
Toll	spatzle	Toll	dorsal	Yes
EGF-R	spitz vein gurken	torpedo (EGF-R)	pointed	Yes
FGF-R	branchless	breathless heartless	Not known	Yes

later in development, however. Toolkit genes with multiple functions during development are termed **pleiotropic**. For instance, individual genes may act in different developmental events in the same tissue over the course of development, in different tissues at the same stage of development, or in different tissues over the entire course of development. The multiple roles played by individual genes can be unmasked by mutations that affect only a subset of a gene's functions, by conditional mutations (such as temperature-sensitive mutations), or by genetic techniques for investigating the requirements for gene function in subsets of cells at various stages of development.

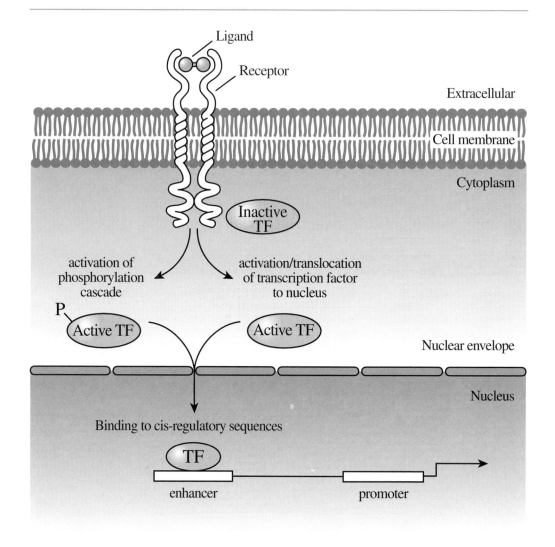

Figure 2.20
A generic signaling pathway

Most signaling pathways operate through similar logic but have different proteins and signal transduction mechanisms. Signaling begins when membrane-bound receptors bind a ligand, leading to the release or activation of associated intracellular proteins. Receptor activation often leads to the modification of inactive transcription factors that are translocated to the cell nucleus, bind to *cis*-regulatory DNA sequences or to DNA-binding proteins, and regulate the level of target gene transcription.

Signaling pathway components are especially pleiotropic. For example, the Wingless pathway plays diverse roles in many tissues throughout *Drosophila* development. It is required early in the ectoderm to organize segment polarity; later in the embryonic ectoderm to direct the formation of the leg and wing imaginal discs; days later in the larval wing field to organize dorsoventral polarity, wing outgrowth, and sensory organ patterning; and finally to organize the polarity of the eye, leg, and other tissues. All components of the Wingless pathway are required in each of these settings, even though the ultimate regulatory and morphogenetic "output" of the pathway is different in each case. In Chapter 3, we will see how transcription factors and signaling pathways achieve their tissue-specific effects.

SHARING OF THE GENETIC TOOLKIT AMONG ANIMALS

Hox *genes*

One of the most exciting and unexpected discoveries that occurred soon after the cloning of the *Hox* genes was the detection of genes with related sequences in all sorts of animals. Using the homeobox as a probe to search for similar sequences in other genomes, researchers isolated *Hox*-related genes from a broad sample of other animals. The similarity between the sequences of the homeodomains of genes isolated from frogs, mice, and humans and the original *Drosophila Hox* sequences was surprisingly extensive given the vast evolutionary distances between these animals. As many as 59 of the 60 amino acid residues were shared between the most similar homeodomains (Fig. 2.21).

An even greater surprise emerged with the physical mapping of vertebrate *Hox* genes. The map revealed that these *Hox* genes occurred in four large, linked complexes and that the order of the *Hox* genes within these complexes followed the order of their most related counterparts in the insect *Hox* complexes (Fig. 2.22). The vertebrate complexes define 13 groups of *Hox* genes, compared with the 8 genes in *Drosophila*, although not every *Hox* gene is represented in each vertebrate complex (see Fig. 2.22). Furthermore, the relative order of expression of vertebrate *Hox* genes along the anteroposterior (rostrocaudal) axis of vertebrate embryos correlates with gene position in each complex (see Fig. 2.22).

With the development of techniques for knocking out gene function in mice, it became possible to analyze the functions of the 39 *Hox* genes in the four mouse *Hox* complexes. This analysis has been complicated by **genetic redundancy**—that is, the expression and function of two or more similar *Hox* genes in overlapping domains. In some cases, loss of function of a specific *Hox* gene causes the homeotic transformation of the identity of particular repeated structures, such as vertebrae, and, in other cases, the loss of particular

```
Fly Dfd          PKRQRTAYTRHQILELEKEFHYNRYLTRRRRIEIAHTLVLSERQIKIWFQNRRMKWKKDN  KLPNTKNVR
AmphiHox4        TKRSRTAYTRQQVLELEKEFHFNRYLTRRRRIEIAHSLGLTERQIKIWFQNRRMKWKKDN  RLPNTKTRS
Mouse HoxB4      PKRSRTAYTRQQVLELEKEFHYNRYLTRRRRVEIAHALCLSERQIKIWFQNRRMKWKKDH  KLPNTKIRS
Human HoxB4      PKRSRTAYTRQQVLELEKEFHYNRYLTRRRRVEIAHALCLSERQIKIWFQNRRMKWKKDH  KLPNTKIRS
Chick HoxB4      PKRSRTAYTRQQVLELEKEFHYNRYLTRRRRVEIAHSLCLSERQIKIWFQNRRMKWKKDH  KLPNTKIRS
Frog HoxB4       AKRSRTAYTRQQVLELEKEFHYNRYLTRRRRVEIAHTLRLSERQIKIWFQNRRMKWKKDH  KLPNTKIKS
Fugu HoxB4       PKRSRTAYTRQQVLELEKEFHYNRYLTRRRRVEIAHTLCLSERQIKIWFQNRRMKWKKDH  KLPNTKVRS
Zebrafish HoxB4  AKRSRTAYTRQQVLELEKEFHYNRYLTRRRRVEIAHTLRLSERQIKIWFQNRRMKWKKDH  KLPNTKIKS
```

Figure 2.21

The similarities of *Drosophila* and vertebrate Hox protein sequences

The sequence of the *Drosophila* Dfd homeodomain and C-terminal flanking region and the sequences of several members of the vertebrate *Hox 4* genes are shown. Note the great sequence similarity between the *Drosophila* and vertebrate proteins, and among the vertebrate Hox proteins.

a

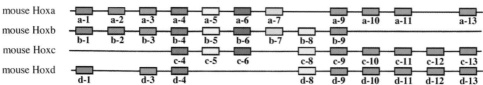

b

Mouse Embryo

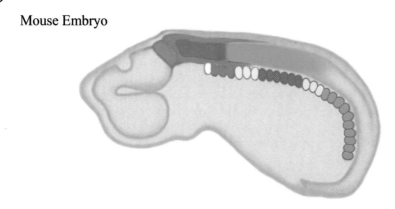

Figure 2.22
***Hox* gene complexes and expression in vertebrates**

(a) In the mouse, four complexes of *Hox* genes, comprising 39 genes in all, occur on four different chromosomes. Not every gene is represented in each complex, however. (b) The *Hox* genes are expressed in distinct rostrocaudal domains of the mouse embryo. Source: Carroll SB. Nature 1995;376:479–485.

organs (Fig. 2.23). Conversely, the expression of *Hox* genes in new sites often causes the reciprocal transformations. Similar results have been obtained in birds, amphibians, and fish, which indicates that in vertebrates, as well as *Drosophila*, *Hox* genes act as region-specific selector genes.

Hox genes also affect the development of unsegmented animals. In the nematode *Caenorhabditis elegans*, for example, *Hox* genes regulate the differentiation of cell types and certain structures along the main body axis. As *Hox* genes have been found on all branches of the metazoan tree and play such important roles in body patterning, we will devote considerable attention to their evolution and function in later chapters.

Field- and cell-type-specific selector genes. The discovery of homologs of *Drosophila Hox* genes in vertebrates and other animal phyla inspired the search for homologs of other *Drosophila* selector and developmental genes. Through cross-hybridization between insect genes and vertebrate genes, families of vertebrate genes were isolated that included homologs of the *Distal-less* (*Dlx*), *tinman* (*Nkx2.5*), and AS-C (MASH) genes. Vertebrate

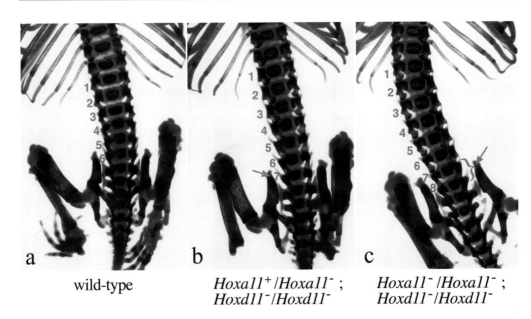

a   b   c

wild-type   *Hoxa11⁺/Hoxa11⁻* ;   *Hoxa11⁻/Hoxa11⁻* ;
                   Hoxd11⁻/Hoxd11⁻   *Hoxd11⁻/Hoxd11⁻*

Figure 2.23
***Hox* genes regulate vertebrate axial morphology**

The morphologies of different regions of the vertebral column are regulated by *Hox* genes. (**a**) In the mouse, normally six lumbar vertebrae arise just anterior to the sacral vertebrae. (**b**) In mice lacking the function of the posteriorly acting *Hoxd11* gene, and possessing one functional copy of the *Hoxa11* gene, seven lumbar vertebrae form and one sacral vertebra is lost. (**c**) In mice lacking both
Hoxa11 and *Hoxd11* function, eight lumbar vertebrae form and two sacral vertebrae are lost. The anterior limit of *Hoxd11* expression
is at the first sacral vertebra. Loss of these *Hox* gene functions transforms the sacral vertebrae into lumbar vertebrae.

Source: Photographs courtesy of Dr. Anne Boulet, HHMI, University of Utah.

homologs of *eyeless* (*Pax6*), *scalloped* (*TEF-1*), *apterous* (*Lmx*), *engrailed* (*En1* and *En2*),
nautilus (*myoD*), and *Dmef* (*mef*), genes are also known.

Thus homologs of most identified field- and cell-type-specific selector genes have been
found in vertebrates. Of even greater interest, however, are the structures and cell types in
which these vertebrate homologs function. For example, the vertebrate *eyeless* homolog,
Pax-6, is involved in the development of the vertebrate eye. Mutations that reduce the
activity of the mouse *Pax-6* gene, called *small eye* (*Sey*), result in loss of eye tissue, including loss of the retina, lens, and cornea in homozygous mutants that lack all *Sey* function
(Fig. 2.24a). Mutations in the human *Pax-6* gene, *Aniridia,* similarly affect eye development. Furthermore, when a version of the *Sey* gene is introduced into and expressed in
flies, it behaves just like the *eyeless* gene in terms of its ability to induce new eye tissue.
The *Pax-6* gene has been found to be associated with eye development across the metazoan tree, despite the differences in the architectures and optic principles of animal eyes.

Similar results have been found for vertebrate homologs of the *Distal-less, tinman,* and
achaete-scute genes. In vertebrates and other phyla, *Distal-less*-related genes are expressed in
an enormous variety of appendages with very distinct morphologies and functions. *Tinman/*

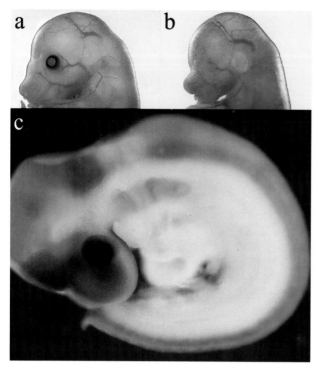

Figure 2.24
The *Pax-6/small eye* gene controls vertebrate eye development

(**a**) Normal late-stage mouse embryo showing the head and developing eye. (**b**) *Sey* mutant embryo lacks the eye entirely. (**c**) In situ hybridization of a *Pax-6/Sey* gene probe to mouse embryos reveals that the gene is expressed throughout the region from which the eye will develop.

Source: Photographs courtesy of Robert Hill and Nadean Brown. Parts a and b from Hill RE, Favor J, Hogan BLM, et al. Nature 1991; 354: 522–525.

NKX2.5 homologs are also expressed in, and required for, the development of the vertebrate heart. Similarly, AS-C homologs are expressed in neural precursors in vertebrates.

Signaling pathways: classical organizers and morphogens

The components of all the major signaling pathways in *Drosophila* also have vertebrate counterparts (see Table 2.2). These pathways operate in many tissues throughout vertebrate development. Importantly, some widely shared signaling proteins have been found to play important roles in the classical organizers defined in vertebrate embryos. For example, searches for molecules with activities associated with the Spemann organizer in the *Xenopus* embryo revealed that several of the secreted signaling proteins have potent inducing activities. One protein, dubbed Chordin because of its activity in inducing dorsal derivatives (the notochord), is a homolog of the *Drosophila* Short gastrulation protein. Both the vertebrate and *Drosophila* proteins interact with members of the TGF-β signaling protein family, and both are involved in dorsoventral axis formation.

Signaling proteins have also been found that account for the activities of the ZPA and AER in the vertebrate limb bud. For example, Sonic hedgehog (Shh), a homolog of the *Drosophila* Hedgehog signaling protein, is expressed in the posterior mesenchyme of the limb bud, precisely where the ZPA is localized (Fig. 2.25). Consistent with Shh carrying out the organizing activity of the ZPA, expression of Shh in the anterior of the limb induces the same ectopic mirror-image duplications of digits as does transplantation of the ZPA. Simi-

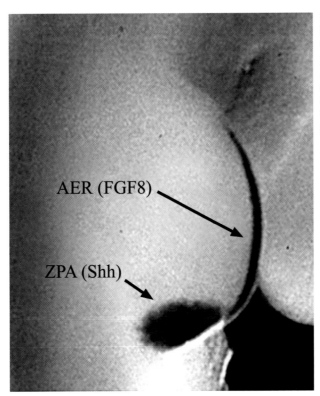

Figure 2.25
Organizers and signaling proteins in the vertebrate limb bud

In situ hybridization of probes for the *FGF8* and *Shh* transcripts reveal that these genes are expressed in regions corresponding to the AER and ZPA, respectively, as defined in transplantation and ablation experiments.

Source: Photograph courtesy of Cliff Tabin.

larly, members of the fibroblast growth factor signaling protein family are expressed in the distal tip of the limb bud in the AER and promote its activity (see Fig. 2.25).

THE TOOLKIT AND ANIMAL DESIGN

The visualization of fields, compartments, and organizers as domains of selector gene expression and as sources of signaling proteins with demonstrable long-range effects on the patterning of embryonic fields has provided concrete molecular evidence of the fundamental roles that these units of organization play in embryonic development. The identification of genes that affect the formation and function of these units of animal body organization represents a first step toward understanding the developmental genetic logic underlying animal design. To understand how an animal is built, we need to understand which mechanisms establish these spatial domains in a growing embryo and how gene products function together to shape animal patterns.

The conservation of the genetic toolkit provokes many developmental and evolutionary questions. How do such different structures as the insect compound eye and the vertebrate lens-type eye develop when their formation is controlled by such similar, even functionally interchangeable genes? And, what does the conservation of genes involved in building anatomically different structures with similar functions tell us about animal ancestors?

We'll confront the mechanistic issues in Chapter 3 and tackle the historical questions in Chapter 4.

SELECTED READINGS

Animal Development—General Texts

Gilbert, S. F. Developmental biology. Sunderland: Sinauer Associates, 1997.

Lawrence, P. The making of a fly. Oxford, UK: Blackwell Scientific Publications, 1992.

Wolpert, L., Beddington, R., Brockes, J., et al. Principles of development. London: Current Biology, 1998.

Organizers, Fields, and Morphogens

Gilbert, S., Opitz, J., Raff, R. Resynthesizing evolutionary and developmental biology. *Develop Biol* 1996; 173:357–372.

Wolpert, L. Positional information and the spatial pattern of cellular differentiation. *J Theoretical Biol* 1969; 25:1–47.

Homeotic Genes in Insects and the Homeobox

Carroll, S. Homeotic genes and the evolution of arthropods and chordates. *Nature* 1995; 376:479–485.

Garcia-Bellido, A. Genetic control of wing disc development in *Drosophila. Ciba Foundation Symp* 1975; 29:161–178.

Kaufman, T. C., Seeger, M. A., Olsen, G. Molecular and genetic organization of the Antennapedia gene complex of *Drosophila melanogaster. Adv Genet* 1990; 27:309–362.

Lewis, E. B. A gene complex controlling segmentation in *Drosophila. Nature* 1978; 276:565–570.

McGinnis, B. A century of homeosis, a decade of homeoboxes. *Genetics* 1994; 137:607–611.

McGinnis, W., Kuziora, M. The molecular architects of body design. *Sci Am* Feb 1994:58–66.

McGinnis, W., Levine, M., Hafen, E., et al. A conserved DNA sequence in homeotic genes of the *Drosophila* Antennapedia and Bithorax Complexes. *Nature* 1984; 308:428–433.

Scott, M. P., Weiner, A. J. Structural relationships among genes that control development: sequence homology between the *Antennapedia, Ultrabithorax,* and *fushi tarazu* loci of *Drosophila. Proc Natl Acad Sci* 1984; 81:4115–4119.

Field-Specific Selector Genes

Cohen, S., Bronner, M., Kuttner, F., et al. *Distal-less* encodes a homeodomain protein required for limb formation in *Drosophila. Nature* 1989; 338:432–434.

Halder, G., Callaerts, P., Gehring, W. Induction of ectopic eyes by targeted expression of the *eyeless* gene in *Drosophila. Science* 1995; 267:1788–1792.

Kim, J., Sebring, A., Esch, J., et al. Integration of positional signals and regulation of wing formation and identity by *Drosophila vestigial* gene. *Nature* 1996; 382:133–138.

Compartment Selector Genes

Diaz-Benjumea, F., Cohen, S. Interaction between dorsal and ventral cells in the imaginal disc directs wing development in *Drosophila. Cell* 1993; 75:741–752.

Lawrence, P., Struhl, G. Morphogens, compartments, and pattern: lessons from *Drosophila? Cell* 1996; 85:951–961.

Morata, G., Lawrence, P. Control of compartment development by the *engrailed* gene in *Drosophila*. *Nature* 1975; 255:614–617.

Cell-Type-Specific Selector Genes

Lassar, A. B., Buskin, J. N., Lockshon, D., et al. MyoD is a sequence-specific DNA binding protein requiring a region of myc homology to bind to the muscle creatine kinase enhancer. *Cell* 1989; 58:823–831.

Michelson, A., Abmayr, S., Bate, M., et al. Expression of a MyoD family member prefigures muscle pattern in *Drosophila* embryos. *Genes Develop* 1990; 4:2086–2097.

Skeath, J., Carroll, S. The achaete-scute complex. Generation of cellular pattern and fate within the *Drosophila* nervous system. *FASEB J* 1994; 8:714–721.

Genetics of Embryonic Axis Formation in Drosophila

Nüsslein-Volhard, C., Wieschaus, E. Mutations affecting segment number and polarity in *Drosophila*. *Nature* 1980; 287:795–801.

St. Johnston, R., Nusslein-Volhard, C. The origin of pattern and polarity in the *Drosophila* embryo. *Cell* 1992; 68:201–219.

Sharing of the Toolkit Among Animals

Duboule, D., Dollé, P. The structural and functional organization of the murine HOX gene family resembles that of *Drosophila* homeotic genes. *EMBO J* 1989; 8:1497–1505.

Gerhart, J., Kirschner, M. Cells, embryos, and evolution. Malden, MA: Blackwell Science, 1997.

Graham, A., Papalopulu, N., Krumlauf, R. The murine and *Drosophila* homeobox gene complexes have common features of organization and expression. *Cell* 1989; 57:367–378.

Krumlauf, R. *Hox* genes in vertebrate development. *Cell* 1994; 78:191–201.

McGinnis, W., Garber, R. L., Wirz, J., et al. A homologous protein-coding sequence in *Drosophila* homoetic genes and its conservation in other metazoans. *Cell* 1984; 37:403–408.

Quiring, R., Walldorf, U., Kloter, U., Gehring, W. J. Homology of the *eyeless* gene of *Drosophila* to the *Small eye* gene in mice and *Aniridia* in humans. *Science* 1994; 265: 785–789.

Riddle, C., Johnson, R., Laufer, E., Tabin, C. Sonic hedgehog mediates the polarizing activity of the ZPA. *Cell* 1993; 75:1401–1416.

CHAPTER 3

Building Animals

The fundamental mystery of development is how a single fertilized egg cell multiplies and differentiates into a complex animal composed of a large number and wide variety of specialized organs, tissues, and cell types. Identification of the genes with the most pronounced effects on development is merely a first step to an understanding of the process. The second challenge is to understand how the activity of these genes unfolds in time and space to control the formation and patterning of animal body plans and body parts. This process is primarily a matter of gene regulation.

Every feature of animal development depends on both serial and parallel gene functions that act within and between regulatory hierarchies. Because evolutionary changes in animal development can arise from changes in the operation of regulatory hierarchies, understanding the genetic logic and molecular circuitry of major regulatory programs is fundamental to understanding the evolution of morphology.

In this chapter, we examine the architecture of developmental regulatory hierarchies at two levels. First, we analyze the genetic logic of the hierarchies that control the formation and patterning of the primary embryonic axes, appendages, organs, and other major features of selected model animals. Second, we analyze the ways in which combinations of regulatory inputs within these hierarchies are integrated at a molecular level by regulatory elements of developmental genes to control their stage- and tissue-specific expression and functions.

GENE REGULATION IN METAZOANS

The evolution of animal bodies composed of different cell types required mechanisms for turning gene expression on and off in specific cells at particular times during development. The temporal and spatial dimensions of animal development, the thousands of genes in animal genomes, the tendency for the same gene product to be utilized in different ways at multiple stages of development, and the packaging of metazoan genes within **chromatin** (a complex of DNA with many associated proteins in cell nuclei) require more complicated regulatory mechanisms to achieve the independent control of individual genes than the mechanisms typically found in unicellular organisms or viruses.

"I should like to work like the archaeologist who pieces together the fragments of a lovely thing which are alone left to him. As he proceeds, fragment by fragment, he is guided by the conviction that these fragments are part of a larger whole which, however, he does not yet know."
—Hans Spemann,
Embryonic
Development and
Induction (1938)

"Several genetic steps seem to be interposed between the reception of a signal, extrinsic to the genome, and its translation first into genetic and later into developmental terms. A hierarchy of genes may be involved in this process, . . ."
—Antonio
Garcia-Bellido
(1975)

Both the protein machinery involved in transcription to produce messenger RNAs (mRNAs) and the structure of genes at the DNA level are more complicated in metazoans than in prokaryotes. Activation of tissue-specific gene expression requires the assembly of protein complexes involving at least three components (Fig. 3.1):

- RNA polymerase II and its associated **general transcription factors**
- Cell-specific, tissue-specific, field-specific, or signal transducer-type **activators**
- **Coactivators** that make contacts between activators and the general transcription machinery and influence the local state of chromatin

Importantly, gene regulation is not all about activation. The spatial and temporal specificity of gene expression depends a great deal on specific **repressors** that act to suppress transcription, often through **corepressors** that affect the local state of chromatin. Many of the genes in the toolkit for animal development are transcriptional activators or repressors.

The general transcription machinery is assembled on DNA sequences called **promoters**, which are found near the site of the start of transcription. Activators and repressors bind to other DNA sequences to affect the transcriptional activity of the gene. Sometimes, these sites may be located in the vicinity of the promoter regions. Often, however, the regulatory proteins that mediate the transcriptional activity of individual developmentally regulated genes bind to specific sites in discrete DNA regions, called *cis*-**regulatory elements** (or *cis*-elements), which can influence the level of transcription by several orders of magnitude, from sites that may be many thousands of base-pairs away from the promoter. The modularity of these elements is illustrated by their ability to regulate transcription of essen-

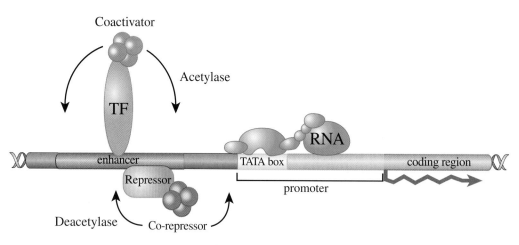

Figure 3.1
Gene regulation in metazoans

The transcription of genes is regulated by transcription factors that bind to *cis*-regulatory DNA sequences. Transcriptional activators (TF) recruit coactivators that open the local chromatin through an associated acetylase activity. Repressors can act through corepressors that have an associated deacetylase activity. RNA polymerase II has many associated transcription factors that are involved in forming an active transcription complex on the promoter, immediately upstream of the site at which transcription is initiated.

tially any gene in *cis*. Such *cis*-elements are usually identified and their function studied by analyzing their activity in regulating a heterologous **reporter gene** (Fig. 3.2). *Cis*-elements may regulate transcription through the looping of distant sequences that brings bound activators and coactivators in contact with the transcription initiation complex at the promoter; they may also act by affecting the state of chromatin (see Fig. 3.1).

The expression of individual genes at particular stages and locations in the developing animal is often controlled by *cis*-elements that are distinct from those controlling gene expression at other times and places. Thus metazoan genes not only consist of the coding sequences for a particular protein, but often possess a modular array of *cis*-elements that act as genetic switches to control gene expression in a variety of different contexts. *The modularity of metazoan cis-regulatory DNA is critical both to the specificity of gene interactions during development and to evolutionary changes in gene expression during the evolution of new morphologies.*

THE ARCHITECTURE OF GENETIC REGULATORY HIERARCHIES

General features and approach

Many genes can affect the same structure, pattern, or process. In many cases, it is impossible to determine the order of gene action solely by examining mutant phenotypes. Nor is it usually clear whether one gene plays a primary role in a process and other genes play subordinate roles. For example, for any two genes *A* and *B* (which based on similar mutant phenotypes appear to control the same process), several regulatory relationships are possible. *A* could affect the expression or function of *B*, *B* could affect *A*, *A* and *B* could affect each other, or *A* and *B* may have no effect on each other. With additional genes, one can quickly see that the number of possible regulatory relationships among a set of genes increases exponentially with the size of the set. Given that genetic screens have often identified multiple loci that have similar developmental effects, it is clear that to decipher the temporal order of action and hierarchical relationships between genes, more information is necessary.

Molecular techniques have revealed that the patterns in which major developmental genes are expressed within various body regions, organs, and cells of the developing animal usually correspond to the structures of the larvae or adult whose formation, pattern, or differentiation are affected by mutations in these genes. This understanding has inspired a whole new approach to embryology and the genetic analysis of developmental regulatory mechanisms. Rather than focus on final physical forms, which are often arrested prematurely and disfigured in lethal mutants, the patterns in which key developmental genes are expressed at different developmental stages can be used as surrogates for the ultimate form. Visualization of gene activity in developing animals provides a much more direct and dynamic (that is, both temporal and spatial) picture of gene function and regulatory interactions than does the inspection of terminal phenotypes (Box 3.1).

The study of model organisms such as *Drosophila* has revealed a few general features of the architecture of genetic regulatory hierarchies underlying developmental programs. First, development is a continuum in which every pattern of gene expression has a preceding causal basis—namely, a previous pattern of gene activities. Second, regulatory information often flows through "nodal points"—key genes that integrate multiple spatial inputs and whose products often control a major feature of the future pattern. For example, the location of organ primordia often requires inputs from anteroposterior and

Developmental Gene

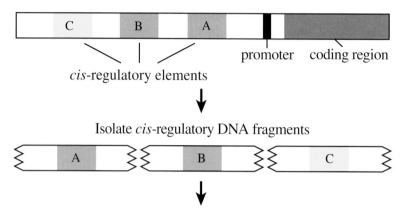

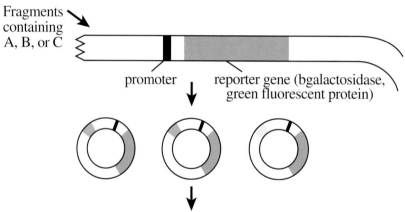

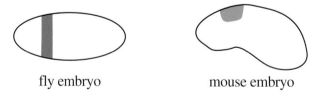

Figure 3.2

Analysis of *cis*-regulatory elements and reporter genes

Animal genes often contain multiple independent elements that control gene expression in different places and/or different times during embryogenesis (e.g., A, B, C). The *cis*-regulatory elements are generally identified through their ability, when placed in *cis* to a heterologous reporter gene and inserted back into a host genome, to control the pattern, timing, and/or level of gene expression. The most frequently employed reporter genes encode either (1) enzymes that can be detected in situ with chromogenic substrates or specific antibodies or (2) proteins that fluoresce in vivo.

Box 3.1   **Gene Logic**

The analysis of gene regulation in animal development has given rise to a number of terms to describe the nature of regulatory interactions. Because the concepts behind these terms are fundamental to understanding genetic logic, it is important to understand the terms and the genetic tests used to define them.

To determine the hierarchical relationships among a set of genes, genetic tests are used to assess the relationships between any two members of the hierarchy. The first question is whether one gene depends on another for proper expression. If the expression of a gene is found to be dependent on another, the first gene is often said to be "downstream" of its "upstream" regulator. The regulatory dependence for expression can either be positive or negative, meaning that a given downstream gene may be dependent on a regulator for its **activation** or **repression**.

Another important distinction is whether the expression of the downstream gene depends on one or many regulators. This issue addresses the criterion of **necessity** versus **sufficiency**. A regulator may be required for expression of a downstream gene (necessary), but may not be capable of activating gene expression on its own (not sufficient).

Finally, as more potential components of a hierarchy are examined, evidence may accumulate regarding whether a particular interaction may be **direct** or **indirect**. If the expression of a downstream gene is affected by multiple upstream genes, then some upstream genes could potentially exert an effect through other upstream genes; hence, those genes have indirect effects on a given downstream gene.

In practice, the architecture of regulatory hierarchies emerges from the detailed analysis of the dependent/independent relationships of individual gene expression patterns with the function of other candidate members of the hierarchy. The combination of genetic tools for altering gene function, molecular techniques for assaying gene expression, and some knowledge of protein function together enable us to establish whether genes of a given phenotypic class control different steps in a single pathway and whether genes of different classes act sequentially or in parallel during development.

---

dorsoventral coordinate systems that are integrated by selector genes, which directly control the formation of these primordia. By focusing on the genetic control of such nodal genes, we can simplify and resolve the regulatory logic of networks that might appear quite complicated at the formal genetic level. Third, many genes, particularly components of signaling pathways and transcription factors, are deployed at several stages of development in distinct spatial patterns and are involved in a variety of regulatory hierarchies.

Regulatory logic—pathways, circuits, batteries, feedback loops, networks, and the connectivity of genes

The analysis of regulatory interactions between genes has given rise to a host of commonly used terms, often applied with somewhat broad meanings, to describe the higher-order relationships and connections between genes. Because the liberal use of these terms can obscure the concepts they are intended to represent, we take a moment here to define their use in this book.

Let's start with a **pathway**. We use this term to describe components that are obligately linked in the transmission of information. Signaling pathways, which may be linear or branched, are composed of components that depend on one or more upstream or downstream components to exert their effects (Fig. 3.3). Multiple structurally related ligands, receptors, or transducers might be able to transmit information in a given pathway.

We define a **circuit** to be larger than a pathway, encompassing additional components that are not obligately linked (see Fig. 3.3). For example, while components of signaling pathways depend on one another to conduct the signal to the cell nucleus, the nature of the pathway output is usually context-dependent. When gene regulation is the output, the target genes of a given individual pathway usually differ between developmental stages and spatial locations in the developing embryo. We use the term "circuit" to describe a regulatory pathway that includes particular target genes. Thus two circuits that employ the same pathway are different if they regulate different target genes.

A group of connected circuits constitutes a regulatory **network**. These circuits may be connected in series or they may consist of parallel circuits that are connected by one or

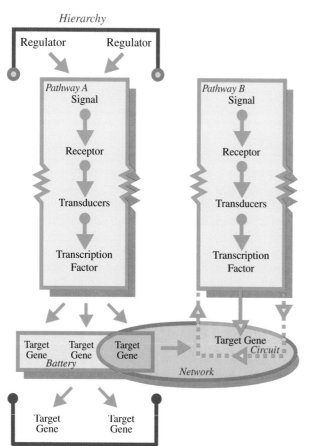

Figure 3.3
Genetic regulatory logic

A model hierarchy, pathway, circuit, battery, and network are depicted. The signaling pathways contain several obligately linked components, but their deployment is controlled by potentially diverse regulators. The target genes they control in regulatory circuits are also diverse and context-specific. Tiers of regulators and targets constitute a hierarchy. Connections between independent regulatory circuits constitute a network.

more links (see Fig. 3.3). Some genes in the network may act earlier than others, or control the activity of many genes, organizing the network into a **hierarchy**. Hierarchies may be vertically organized, comprising many tiers of genes that control lower tiers of genes (see Fig. 3.3). Alternatively, they may be organized more horizontally, with single genes directly controlling the expression of a large group or **battery** of target genes (see Fig. 3.3).

Model regulatory hierarchies and the key genetic switches that operate them

The regulatory mechanisms that underlie a large number of developmental processes in the several model species belonging to a few different phyla have been analyzed in recent years. In this book, we concentrate primarily on insect and vertebrate examples because they illustrate general mechanisms of the control of toolkit gene expression and function during development. Importantly, knowledge derived from these model systems forms the foundation of the comparative approaches to the evolution of morphology described in subsequent chapters.

Our discussion focuses on the regulatory hierarchies that control the sequential generation of spatial coordinate systems that unfolds during embryogenesis—from the generation of the major body axes, to the formation of primary and secondary fields, to the patterning of individual fields. We examine at the molecular level how organizers function within embryos and fields, how gradients of morphogens are interpreted within fields, how selector genes regulate the formation and identity of individual fields, and how *cis*-regulatory elements integrate combinations of regulatory inputs to control complex spatial patterns of toolkit gene expression.

THE INSECT BODY PLAN

The genetic regulatory hierarchies that establish the major features of the *Drosophila melanogaster* body plan are of interest for two reasons: (1) because they are among the best understood in any animal and illustrate potentially general principles, and (2) because they provide a basis for comparison between *Drosophila* and other insect and arthropods. The structure of these regulatory hierarchies is critical to mechanisms of arthropod body plan evolution. The major features of the insect body plan that have both developmental and evolutionary significance include the following:

- Segmentation
- The organization of segments into distinct regions forming the head, thorax, and abdomen
- The identities of paired, jointed appendages on different head and thoracic segments
- The limbless abdomen
- The two pairs of dorsal flight appendages

Fruit flies are **holometabolous** insects, meaning that they undergo a complete metamorphosis between their larval and adult forms during the pupal stage. The larval body

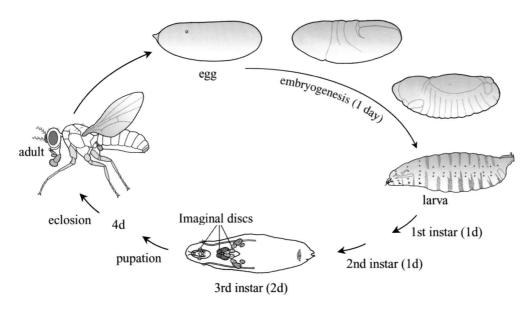

Figure 3.4
The *Drosophila* life cycle

In *Drosophila*, the development of the segmented, motile larva from a fertilized egg takes about one day. After two larval instars, the imaginal discs of the late third larval instar that will give rise to adult tissues are well developed. Morphogenesis and differentiation of adult tissues take place during the pupal stage, before the adult emerges (eclosion).

plan differs from the adult body plan (for a view of the fly life cycle, see Fig. 3.4). Here we examine regulatory mechanisms involved in both larval and adult patterning.

From egg to segments: the anteroposterior coordinate system

In just one day following fertilization, the *Drosophila* embryo develops from a single nucleus in a huge yolk-filled egg about 0.5 mm long into a highly organized segmented, motile, feeding larva with a complex nervous system and the future adult tissues growing as imaginal structures within it. Genetic screens identified five tiers of regulatory genes involved in organizing body pattern along the primary anteroposterior (A/P) axis of the developing embryo. As discussed in Chapter 2, the maternal effect, gap, pair-rule, segment polarity, and homeotic genes have large and distinct effects on the patterning of the A/P axis.

Genetic and molecular analyses of the regulatory interactions within and among these five tiers of genes have elucidated both the general logic of and many specific molecular interactions that operate within the A/P patterning hierarchy. These molecular mechanisms, in turn, illustrate several important general concepts about the function and transcriptional regulation of pattern-regulating genes during animal development.

Figure 3.5 shows the basic outline of the A/P axis regulatory hierarchy. The generation of the periodic, segmental organization of the larva from an initially aperiodic egg involves regulatory interactions between gene products translated from maternal mRNAs deposited in the egg, transcriptional activation of zygotic genes by certain maternal activators, and combinatorial action of segmentation gene products to refine the expression patterns of many zygotic segmentation genes into iterated domains. This hierarchy has four key features:

1. Generation of gradients of maternal transcription factor proteins
2. Transcriptional activation of and cross-regulation by the gap genes

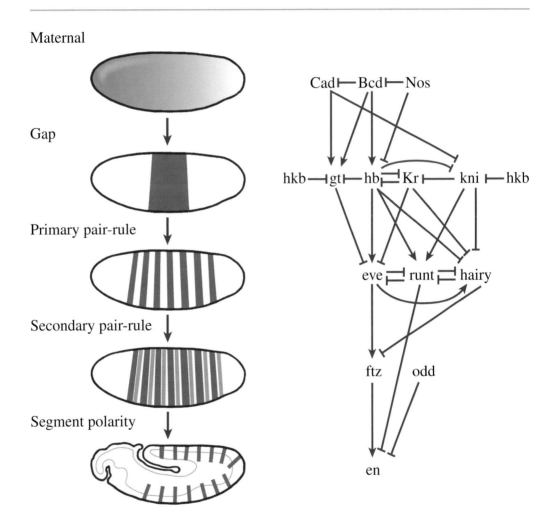

Figure 3.5
The segmentation genetic regulatory hierarchy

(**left**) The expression patterns of five classes of anteroposterior axis patterning genes are depicted in embryos at different stages. (**right**) Selected members of these classes are shown and the regulatory interactions between these genes are indicated. An arrow indicates a positive regulatory interaction; a line crossed at its end indicates a negative, repressive regulatory relationship.

3. Transcriptional regulation of individual pair-rule gene stripes by combinations of maternal and gap proteins

4. Regulation of segment polarity gene expression by pair-rule and segment polarity proteins

Generation of maternal transcription factor gradients. The Bicoid (Bcd) and Caudal (Cad) proteins are homeodomain-type transcription factors, and the Hunchback (Hb) protein is a zinc-finger-type transcription factor. The mRNAs encoding these proteins are deposited in the egg and translated during early embryonic development. Because the early embryo is a **syncytium**, lacking any cell membranes that would impede the diffusion of protein molecules, these transcription factors can move freely throughout the cytoplasm. The concentration of each protein is graded along the A/P axis of the embryo (Fig. 3.6). These maternal gradients are important because several downstream segmentation genes are regulated by Bcd, Cad, and Hb (depending on their concentrations). Early gene regulation in the fly embryo provides a general model for the concentration-dependent control of gene expression by gradients of regulatory proteins.

In *Drosophila,* these three maternal gradients are formed by different mechanisms. The Bicoid gradient results from the localization of maternal mRNA at the anterior pole of the egg. Translation of this mRNA and diffusion of the Bicoid protein toward the posterior creates a concentration gradient. The Hunchback gradient is generated by selective inhibition of the translation of the ubiquitous *hb* mRNA in posterior regions (see Fig. 3.6b). This inhibition is regulated by the product of the *nanos* gene, which binds to the *hb* mRNA. The Nanos (Nos) protein is in a posterior-to-anterior concentration gradient, generated by the localization of its mRNA to the posterior end of the egg. The *caudal* mRNA, like *hunchback*, is distributed evenly throughout the egg, although selective inhibition of its translation by the Bicoid protein generates a posterior-to-anterior Caudal protein gradient (see Fig. 3.6b).

Transcriptional activation of and cross-regulation by gap genes. The gap genes are the first zygotic genes to be expressed in discrete regions along the A/P axis. The Bcd, Hb, and Cad proteins are involved in the initial regulation of these genes. One key regulatory interaction is the activation of the second phase of *hunchback* expression by the Bicoid protein in the anterior half of the embryo. This activation occurs through a direct interaction of the Bicoid protein with a *cis*-regulatory element of the *hb* gene. DNA sequences 5′ to the *hb* promoter contain several binding sites for the Bicoid protein that are necessary for the Bcd-dependent activation of the *hb* gap domain (see Fig. 3.6c). Bcd binds to these sites **cooperatively**—that is, the binding of one Bcd protein molecule to one site facilitates the binding of other Bcd molecules to nearby sites. More than one occupied Bcd site is necessary to generate the sharp boundary of *hb* transcription, suggesting that this *cis*-regulatory element requires a threshold concentration of the Bcd protein to occupy multiple binding sites before it can activate *hb* expression. In this way, the *hb* gene "reads" or "interprets" the Bcd gradient.

Different gap genes are deployed in distinct domains in the embryo, depending on the levels of the Bcd, Cad, and Hb proteins. Each gene possesses *cis*-regulatory elements that contain different arrangements of binding sites with different affinities for these transcription factors. Consequently, gap genes are activated at different positions along the A/P axis. Their expression domains are further refined through mutually repressive interactions

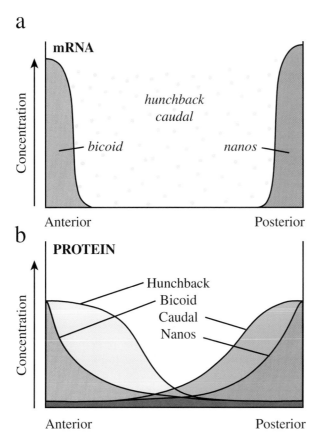

a

mRNA

hunchback
caudal

bicoid   nanos

Anterior   Posterior

b

PROTEIN

Hunchback
Bicoid
Caudal
Nanos

Anterior   Posterior

c

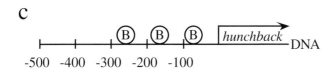

-500 -400 -300 -200 -100

Figure 3.6
Maternal gradients and gene activation

(**a**) The *bicoid* and *nanos* mRNAs are localized to the anterior and posterior poles, respectively, whereas the *hunchback* and *caudal* mRNAs are found throughout the syncytial embryo. (**b**) Diffusion of the Bicoid and Nanos proteins leads to the formation of concentration gradients. Bicoid activates *hunchback* transcription and represses Caudal translation; Nanos represses *hunchback* translation, leading to graded distributions of all four proteins. (**c**) The *hunchback* gene contains several binding sites for the Bicoid protein upstream of the promoter. Occupancy of these sites leads to activation of the *hunchback* gene in the anterior half of the embryo.

Source: Modified from Gilbert S. Developmental biology, 5th ed. Sunderland: Sinauer Associates, 1997.

between gap proteins (all of which are transcription factors) and the *cis*-regulatory elements of gap genes expressed in adjacent domains.

Initiation of periodic pair-rule gene expression. The aperiodic maternal and gap protein expression patterns establish the periodic, seven-striped patterns of certain pair-rule genes, called "primary" pair-rule genes. The way in which the pattern of pair-rule stripes is created by aperiodically distributed regulators had been one of the major mysteries of *Drosophila* segmentation. The solution to this puzzle highlights one of the most important concepts in the spatial regulation of gene expression in developing animals—namely, the independent control of distinct *cis*-regulatory elements in a single gene.

The key discovery was that the seven stripes that make up the expression patterns of the primary pair-rule genes *hairy* and *even-skipped* are controlled independently. That is, different *cis*-regulatory elements control the expression of different stripes. The seven-striped pattern of *eve* expression, for example, represents the sum of seven sets of regulatory inputs into seven separate *cis*-elements of the gene (Fig. 3.7). Detailed analysis of individual stripe *cis*-regulatory elements revealed that the position of a "simple" stripe was controlled by no fewer than four regulatory proteins. For example, for the second *eve* stripe, Bcd and Hb act as broadly distributed activators, and the boundaries of the stripe are sharpened via repression by the Giant and Krüppel gap proteins (see Fig. 3.7). This roughly 700 bp element contains multiple binding sites for each activator and repressor. The *cis*-regulatory element acts as a genetic switch, integrating a host of regulatory protein activities to regulate *eve* expression in one three- to four-cell-wide stripe in the embryo.

Other *eve* and *hairy* stripe elements are regulated by different combinations of maternal activators and gap proteins found at other positions in the embryo. The periodic expression of these primary pair-rule proteins, in turn, regulates the periodic expression of downstream "secondary" pair-rule genes.

Regulation of segment polarity genes by pair-rule proteins. The 14-stripe patterns of several segment polarity genes (corresponding to 14 segments) are initiated by the pair-rule regulatory proteins, then subsequently refined and maintained by the segment polarity gene products themselves. Pair-rule proteins regulate alternating (7 out of 14) segment polarity stripes. For example, Ftz acts through a *cis*-regulatory element in the *engrailed* gene to help activate the even-numbered *en* stripes. The Odd-skipped (Odd) pair-rule protein is expressed slightly out of phase with the Ftz protein and represses expression of the *en* gene. In this manner, narrow *en* stripes are activated in cells that express Ftz but do not express Odd.

After the pair-rule proteins position the initial expression domains of segment polarity genes in the growing (now cellularized) embryo, interactions among the segment polarity proteins and genes refine and maintain these patterns. These regulatory circuits are important because they maintain boundaries between compartments and segments; they also regulate the expression of other genes that control the morphogenesis of polarity. The pathways that operate within these circuits include some of the most widely deployed pathways in animal development. Much has been learned about their general mechanisms through the study of segment polarity in *Drosophila*.

Two major signaling pathways regulate the maintenance and elaboration of polarity within *Drosophila* segments: the Wingless and Hedgehog pathways. The Hh protein is produced in all posterior cells in each segment, which also express *engrailed*. Hh signals anterior cells through the Ptc receptor and several transducers (Fig. 3.8), to activate *wingless* expression in adjacent anterior cells. Wg, in turn, signals back to posterior cells through the Wingless receptors (D-fz) and various signal transducers (see Table 2.2), to maintain *engrailed* expression, *Hh* expression, and the continuity of the regulatory circuit (see Fig. 3.8). Signaling between cell populations to initiate and maintain gene expression patterns is a general feature of cellular fields. Indeed, these signaling molecules play a wide variety of roles in animal development.

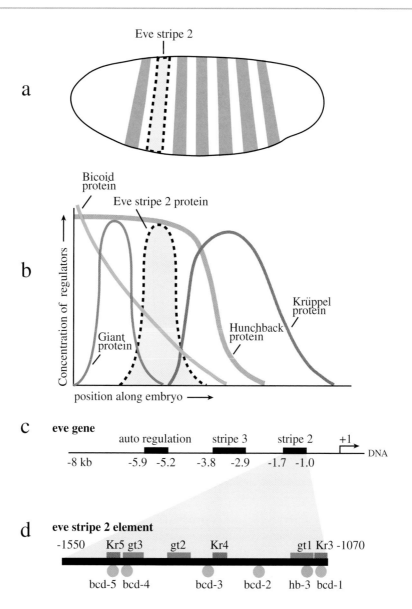

Figure 3.7

Regulation of a pair-rule stripe: combinatorial control of an independent *cis*-regulatory element

The regulation of the *even-skipped cis*-regulatory element controls the formation of the second stripe in the early embryo. (**a**) The stripe 2 element controls just one of seven stripes of *eve* expression. (**b**) The stripe forms within the domain of the Bicoid and Hunchback proteins and at the edge of the Giant and Krüppel gap protein domains. The former are activators, and the latter are repressors, of *eve* stripe 2 expression. (**c**) The *eve* stripe 2 element spans from about 1.7 to 1.0 kilobases upstream of the *eve* transcription unit. (**d**) Within this element, several binding sites for each regulator exist. The net output of the combination of activators and repressors is expression of the narrow *eve* stripe.

Source: Modified from Gerhart J, Kirschner M. Cells, embryos and evolution. Malden, MA: Blackwell Science, 1997.

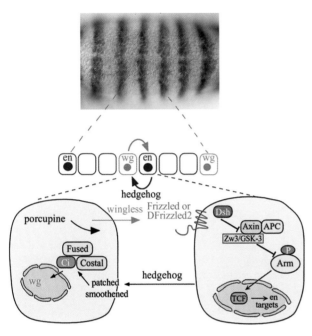

Figure 3.8
The regulation of segment polarity gene expression in *Drosophila* segments

The maintenance of segment polarity gene expression in specific domains within each segment is controlled by signaling interactions between cells. (**top**) Embryo double-labeled to reveal *wingless* (black) and *engrailed* (brown) expression in stripes of adjacent cells in each segment. (**bottom**) Hedgehog signaling from Engrailed-expressing cells (**right**) to cells anterior (**left**) induces *wingless* expression through the members of the Hedgehog pathway (Patched/Smoothened/Costal/Fused/Ci). Wingless, in turn, signals back to posterior cells through components of the Wingless pathway (Frizzled/DFrizzled2/Dsh/Zw3/Arm/TCF) to maintain *engrailed* expression.

Source: Figure parts courtesy of Röel Nusse and Nipam Patel.

General lessons from the segmentation genes

The A/P segmentation hierarchy illustrates five general concepts concerning the spatial regulation of gene expression in animal development:

1. The **concentration-dependent response** of genes to graded inputs is illustrated by the regulation of gap target genes by the Bicoid protein. Both threshold and graded responses to inducers are major themes of gene regulation in cellular fields as well.

2. The **action of both activators and repressors** determines all of the gap and many of the pair-rule and segment polarity gene expression patterns. The refinement of gene expression patterns from those covering most of the embryo (maternal proteins), to 15- to 20-cell-wide regions (gap proteins), to 3- to 4-cell-wide stripes (pair-rule proteins), and ultimately to 1- to 2-cell-wide stripes (segment polarity proteins) depends on activators that define potential areas of gene activation and repressors that restrict the areas in which target genes are expressed. The spatial repression of gene expression to carve ever-finer patterns out of larger domains is a major regulatory theme in cellular fields.

3. Many genes are regulated by two or more activators or repressors. **Combinatorial control** mechanisms impose greater specificity and allow for a greater diversity of spatial patterns.

4. **Multiple independent *cis*-regulatory elements** regulate the expression of many segmentation genes. Individual elements control individual domains of gene expres-

sion (for example, each pair-rule stripe). Furthermore, many segmentation genes are also expressed in the developing nervous system in unique patterns that are controlled by other discrete *cis*-regulatory elements. The utilization of multiple independent elements controlling different spatial domains of gene expression is a general theme of developmental gene regulation.

5. The sequential activation of gap, pair-rule, and segment polarity genes, with each tier of genes being dependent on the preceding tier, constitutes a **regulatory hierarchy**. The domino-like activation of genes in sequence assures the proper temporal deployment of developmental genes.

The dorsoventral axis coordinate system

Like the A/P axis, the dorsoventral (D/V) axis of the *Drosophila* embryo is subdivided into a series of domains that give rise to different tissues. Ventral-most cells will form mesoderm, more ventrolateral cells will give rise to the neuroectoderm, lateral regions generate the dorsal epidermis, and dorsal cells produce an extraembryonic structure, the amnioserosa. Although subdivision of the D/V axis is accomplished by an entirely different set of genes from those that regulate the A/P axis, both transformations occur through similar transcriptional regulatory mechanisms. Combinations of activators and repressors, including some that work in a concentration-dependent manner, act on discrete *cis*-regulatory elements to carve out the spatial boundaries of downstream pattern-regulating genes.

A maternal transcription factor gradient also organizes the D/V axis. This gradient is established by regulating the nuclear localization of the Dorsal (Dl) protein. The highest concentrations of the nuclear Dorsal protein are found in ventral cells, lower levels occur in ventrolateral and lateral regions, and no nuclear Dorsal protein is found in dorsal-most regions (Fig. 3.9a,c). The protein is required in the ventral region to induce ventral tissues.

Several zygotic genes are regulated by different threshold responses to Dorsal protein concentration along the D/V axis (Fig. 3.9b). The response of genes to the Dorsal gradient depends on the number and affinity of binding sites for the Dorsal protein within the *cis*-regulatory elements of target genes. Low-affinity sites are occupied only at high concentrations of Dl nuclear protein, which are found in ventral-most cells. For example, the *twist* and *snail* response elements contain low-affinity sites and are activated in ventral-most cells. In contrast, high-affinity sites can be occupied at low concentrations of Dorsal. The *rhomboid* gene *cis*-element, for example, contains high-affinity sites and is activated by lower concentrations of Dorsal in lateral regions. This protein also acts as a repressor to prevent the expression of genes such as *zen* and *dpp* in ventral and lateral regions of the embryo.

The sharpness of the expression boundaries of the zygotic D/V patterning genes is refined by further regulatory interactions. The *snail* (*sna*) gene enhancer, which is bound by Dl, also requires the product of the *twist* (*twi*) gene to be activated. Synergistic interactions between Dl and Twi ensure a sharp on/off border of *sna* expression, which coincides with the boundary between and regulates the differentiation of the mesoderm from the neuroectoderm (see Fig. 3.9c). The *rhomboid* (*rho*) -element is repressed by the Snail protein, which eliminates *rho* expression from the mesoderm and restricts it to the neuroectoderm (see Fig. 3.9c). A cascade of subsequent interactions, involving primarily the activity of the Dpp signaling molecule, further subdivides the dorsal region of the embryo.

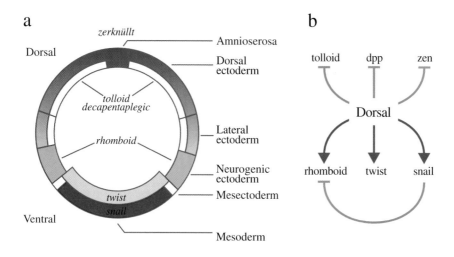

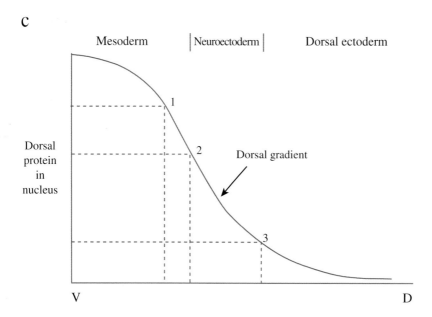

Figure 3.9

The Dorsal gradient and the dorsoventral genetic regulatory hierarchy

(a) Different zygotic genes are expressed in different spatial domains along the D/V axis of the embryo. Some of these domains correspond to populations of cells that will give rise to distinct regions of the embryo (such as the mesoderm and neurogenic ectoderm). (b) The battery of genes regulated by Dorsal protein. (c) Dorsal regulation of target genes is dependent on threshold concentrations of protein. The concentration of nuclear Dorsal protein is highest in ventral (V) cells and lowest in dorsal (D) cells. Certain target genes are activated at different concentrations of Dorsal (thresholds 1, 2, and 3).

Source: Adapted from Gilbert S. Developmental biology, 5th ed. Sunderland: Sinauer Associates, 1997; Jiang J, Levine M. Cell 1993;72:741–752.

The **Hox** *ground plan*

In *Drosophila* and many other animals, the spatial and temporal expression patterns of individual *Hox* genes are much more complicated than those of most segmentation or D/V axis-patterning genes. The number of *cis*-elements and *trans*-acting regulators for each gene is therefore considerably greater for the *Hox* genes. The regulation of *Hox* gene expression has the following major features:

- Activation in a broad domain, usually consisting of two or more parasegments
- Modulation of the levels of expression within this domain
- Expression in all three germ layers, often in patterns that are slightly out of register between each germ layer
- Dynamic changes in the level and domains of expression throughout the subsequent course of development

Numerous transcription factors act on *Drosophila Hox* genes; likewise, many *cis*-acting regulatory elements control the various features of *Hox* expression. In the embryonic ectoderm, for example, most *Hox* genes are regulated by as many as six classes of regulators. All three classes of zygotic segmentation genes are involved in *Hox* regulation. Gap proteins regulate the broad initial domains of *Hox* transcription, selected pair-rule proteins are involved in setting the initial spatial register of certain *Hox* gene domains, and segment polarity proteins, such as the Engrailed protein, help provide intrasegmental modulation of *Hox* gene expression levels. A fourth level of control involves other Hox proteins—the posterior boundaries and levels of expression of some *Hox* genes are regulated by more posteriorly acting Hox proteins. A fifth level of control involves autoregulatory feedback mechanisms that operate to maintain *Hox* expression domains. Finally, a large group of chromatin-associated proteins, termed the Polycomb group and Trithorax group of proteins, act to maintain the repressed or activated transcriptional state, respectively, of many *Hox* genes (Fig. 3.10).

The *Ultrabithorax* gene provides a good example of the constellation of *cis*-regulatory elements that control *Hox* gene expression. Different *cis*-elements of *Ubx* control expression in particular parasegmental domains, in a manner analogous to individual pair-rule stripe elements (Fig. 3.11). These elements contain binding sites for several activators and

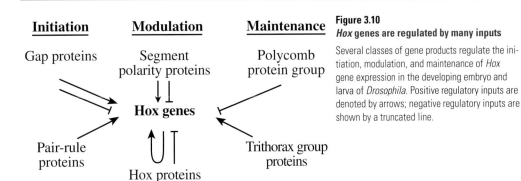

Initiation   **Modulation**   **Maintenance**

Gap proteins   Segment polarity proteins   Polycomb protein group

Hox genes

Pair-rule proteins   Hox proteins   Trithorax group proteins

Figure 3.10
Hox genes are regulated by many inputs

Several classes of gene products regulate the initiation, modulation, and maintenance of *Hox* gene expression in the developing embryo and larva of *Drosophila*. Positive regulatory inputs are denoted by arrows; negative regulatory inputs are shown by a truncated line.

a

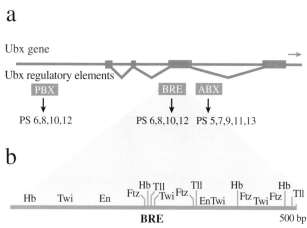

Ubx gene

Ubx regulatory elements

PBX

↓

PS 6,8,10,12

BRE   ABX

↓   ↓

PS 6,8,10,12   PS 5,7,9,11,13

b

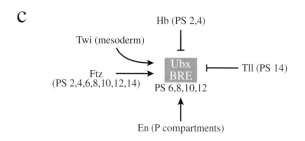

| Hb | Twi | En | Ftz | Hb Tll | Twi Ftz | Tll | En Twi | Hb | Ftz Twi Ftz | Hb | Tll |

BRE   500 bp

c

Hb (PS 2,4)

Twi (mesoderm)

Ftz
(PS 2,4,6,8,10,12,14)

Ubx
BRE
PS 6,8,10,12

Tll (PS 14)

En (P compartments)

Figure 3.11
The *cis*-regulatory elements of the *Ubx* gene

(**a**) Three elements that regulate *Ubx* expression in the embryo are shown. Each element controls expression in selected subsets of parasegments. (**b**) The BRE element contains binding sites for several segmentation proteins (Ftz, Hb, En, Tll) and a dorsoventral patterning protein (Twi) within a 500 bp span. (**c**) Expression driven by the BRE element in parasegments 6, 8, 10, and 12 is the net output of the various positive and negative regulatory inputs shown in this figure.

repressors. The pattern of *Ubx* expression controlled by these elements is the net output from numerous positive and negative inputs. Other *Ubx* elements control the level of expression within parts of parasegments, in appendage fields later in development, and in other germ layers such as the mesoderm (see Fig. 3.11).

Secondary fields: integrating the anteroposterior and dorsoventral coordinate systems

The anteroposterior and dorsoventral axis regulatory genes establish a two-dimensional coordinate system over the entire embryo and within each segment. The *Hox* ground plan is superimposed on this coordinate system and provides the regulatory information for differentiating the events that take place within different segments.

Once the two coordinate systems and *Hox* ground plan are deployed, the development of secondary fields begins in the *Drosophila* embryo. Neural precursors and the primordia for structures such as the heart, salivary glands, and imaginal discs (which will later give rise to the adult appendages and body wall) form at discrete A/P and D/V coordinates within particular segments. The positions of these structures in the embryo are specified in a manner analogous to the way a geographic position is defined on the globe through the combination of latitude and longitude.

Specification of organ precursor position, number, and identity requires integration of regulatory inputs from both coordinate systems, the *Hox* genes, and sometimes other regulators. As described in Chapter 2, the development of various precursors and organ fields is controlled by specific selector genes. The activation of these genes is often the initial step in the specification of these structures. The regulation of these field- and cell-type-specific selector genes reflects the molecular output of the integration of the various spatial regulatory systems. The analysis of their regulation reveals a great deal about the genetic logic underlying early organogenesis. There are several well-studied examples of the integration of axial inputs in the specification of embryonic organs and secondary fields.

The limb fields. The ventral limb fields (legs, mouthparts, and others) and dorsal flight appendage fields (wing, haltere) arise from small populations of approximately 20 cells that are initially specified during mid-embryogenesis. The first sign of the specification of these primordia is the activation of regulatory genes within them. The expression of the *Distal-less* gene marks the development of the ventral limb primordia. *Dll* is expressed initially in small clusters of cells in the ventrolateral ectoderm in many of the head segments and in each thoracic segment. The ventrolateral and anteroposterior positions of these clusters of cells are specified by signals that organize patterning along the dorsoventral and anteroposterior axes of each segment.

Both positive and negative regulatory inputs are used to position the clusters. Along the A/P axis, the Wingless signaling protein is a positive regulator of *Dll* expression. *Dll* is not induced everywhere that the Wg signal is present, however. Instead, *Dll* expression is restricted to a ventrolateral position by negative regulation in dorsal cells by the Dpp signal and in ventral cells by a signal acting through the EGF receptor pathway (Fig. 3.12).

All three signals—and hence the proper coordinates for *Dll* expression—are present in each trunk segment, yet *Dll* is not expressed in the abdominal segments. This situation occurs because *Hox* regulation of *Dll* is superimposed on the segmentally reiterated coordinate system to prevent activation in the abdomen. The Ubx and abd-A proteins bind directly to and repress a *cis*-regulatory element of the *Dll* gene that controls *Dll* expression in the embryonic thoracic segments (see Fig. 3.12).

The *Dll* gene possesses many *cis*-regulatory elements that function in limb primordia in parts of the embryonic head and/or thorax. Some of these elements are regulated by other *Hox* genes; others appear to be redundant. The existence of several elements allows for the independent control of the timing, level, and pattern of *Dll* expression in different segments and at different stages of development.

The wing primordia. The formation of the wing primordia is also regulated by intrasegmental signals and repressed in specific segments by Hox proteins. Cells that form the wing primordia migrate out of the limb field and are specified later and at a more dorsal position than are the limb primordia. Unlike the limb primordia, the wing primordia form in a region characterized by high levels of Dpp signaling (see Fig. 3.12). Primordia formation is selectively repressed in the first thoracic and all abdominal segments by the Scr, Ubx, and Abd-A proteins (see Fig. 3.12). We will have more to say about the role of *Hox* genes in the evolution of insect limb and wing number in Chapter 5.

The salivary gland. The salivary gland primordia form within the ventral epidermis of parasegment 2 in the developing embryo. The positioning of this organ and its restriction to parasegment 2 follow a similar logic as we have described for the appendage primor-

DROSOPHILA EMBRYO

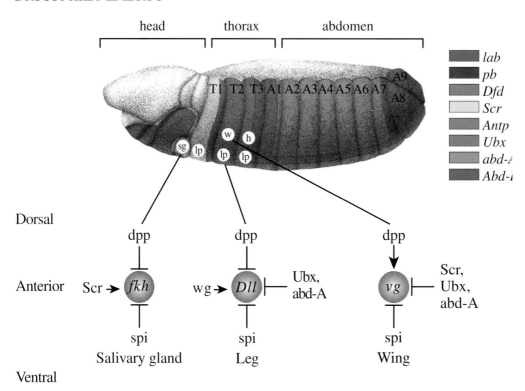

Figure 3.12
Regulation of organ primordia number and position by intercellular signals and Hox proteins
The salivary gland (sg), limb (lp), wing (w), and haltere (h) primordia arise in specific segments at particular positions along the D/V axis. The formation of these primordia is marked by expression of specific genes (e.g., *fkh*, *Dll*, *vg*). The position of each primordium and the expression of marker genes are regulated by signals along the D/V axis of each segment. Specific Hox proteins regulate the segment-specific expression of marker genes and primordium formation.

dia. Expression of the transcription factor Forkhead (Fkh) is an early event in salivary gland development. The regulation of the *fkh* gene parallels the regulation of salivary gland organogenesis. Although the *fkh* gene and salivary gland development are activated by the Hox protein Scr in parasegment 2, they are limited to the ventral epidermis by negative regulation, from the Dpp signal dorsally, and the EGF receptor pathway more ventrally (see Fig. 3.12).

Neural and muscle precursors. The patterns of embryonic neural and muscle precursors in the ectoderm and mesoderm, respectively, are also generated by combinatorial inputs of A/P and D/V signals and modified by Hox regulatory inputs. Rows of proneural clusters that will give rise to segmentally iterated patterns of neuroblasts in the early embryo are marked by the expression of Achaete-Scute Complex genes, whose regulation is a nodal point for the patterning of the central and peripheral nervous system. In the AS-C, *cis*-regulatory elements integrate inputs from segmentation proteins and dorsoventral-axis patterning proteins that establish proneural clusters. Later in embryogenesis, segment-specific patterns of peripheral neural precursors arise from combinatorial regulation of AS-C genes by segmentation, dorsoventral, and Hox regulatory inputs.

The AS-C contains many *cis*-regulatory elements that control discrete subpatterns of AS-C gene expression. Muscle precursors, marked by the expression of the *nautilus* gene, are positioned by combinatorial signals within the mesoderm as well as by Hox inputs that regulate segment-specific muscle patterns.

Patterning within secondary fields: organizing signals and selector genes

The intrasegmental coordinate systems in the embryo and the *Hox* ground plan control the initial position, size, and number of developing *Drosophila* fields. As these fields grow and begin to differentiate, regulatory hierarchies within these fields establish new coordinate systems and organizers that control the formation and morphogenesis of the final structure.

Perhaps the best-studied genetic regulatory program of all secondary fields is that of the *Drosophila* wing. This wing develops from a portion of the larval imaginal disc that will also give rise to the body wall of the second thoracic segment. The wing disc grows from roughly 30 to 40 cells in the first instar larva to approximately 50,000 cells in the third instar larva before morphogenesis transforms it into the adult wing and thoracic body wall. Coordinate systems and selector genes operate within the wing field to control the position, number, and differentiation of the various pattern elements.

The regulatory logic and several of the molecular mechanisms involved in the hierarchies that control wing formation and patterning have been described in sufficient detail that some general concepts regarding the patterning of secondary fields have emerged. These concepts, which apply to other insect appendages as well as to vertebrate limbs, involve regulatory mechanisms that are similar to those operating to position and regulate the formation of pattern elements in other cellular fields. Because the number, position, size, and morphology of pattern elements within secondary fields are important aspects of morphological diversity, understanding these regulatory mechanisms is crucial to developing a picture of the evolution of gene expression and morphology within a field.

The development of secondary fields from primordia established in the embryo involves three major processes:

1. The generation of new coordinate systems during the growth of the field

2. The placement and specification of pattern elements within the field

3. The differentiation of field identity from other, serially homologous fields

The genetic regulatory mechanisms governing these processes involve the integration of combinatorial inputs by *cis*-regulatory elements, similar to those described earlier for the generation of body axes and the initial specification of the fields. One set of regulators includes short- and long-range signaling proteins (morphogens) that determine the area within a field in which a given gene is activated or repressed. The sources of these signals are usually oriented with respect to one patterning axis of the field (anteroposterior, dorsoventral, or proximodistal). A second regulator controls the field- or cell-type-specific response to these signals, such as a selector gene. The identity of serially homologous fields is determined by another layer of regulation superimposed on the signaling inputs and field-specific selector genes by *Hox* genes or analogous regulators that are expressed in only one serial homolog. The integration of signaling, field-specific selector, and *Hox* inputs through *cis*-regulatory elements of target genes controls gene expression patterns in a particular field.

In the *Drosophila* wing and haltere (the wing's serial homolog), the identity and function of the major signaling (Dpp, Hh, Wg, N), wing-specific selector (Vg, Sd), and *Hox* (Ubx) inputs in the wing patterning hierarchies have been studied extensively. Three coordinate systems operate in the wing field:

- The first coordinate system is involved with patterning along the A/P axis.

- The second coordinate system acts along the D/V axis.

- The third coordinate system, the proximodistal (P/D) axis, affects the integration of these inputs and the specification of the entire wing field.

Next, we examine the genetic regulatory hierarchies that govern the formation and operating of these coordinate systems and see how they control gene expression and wing pattern.

The anteroposterior coordinate system. The A/P coordinate system encompasses the sequential organizing activities of the Engrailed, Hedgehog, and Dpp proteins. Patterning along the A/P axis of the wing is controlled by a hierarchy involving compartment selector genes and multiple signaling pathways.

Posterior cells are segregated from anterior cells by a lineage restriction imposed by the *engrailed* gene, which is expressed in all posterior cells but not in anterior cells. This lineage restriction creates a smooth boundary between the anterior and posterior compartments, across which important inductive events take place. Due to the regulatory activity of *engrailed*, cells in the posterior (but not the anterior) compartment express the short-range signaling molecule Hedgehog (Fig. 3.13a). Cells in the anterior compartment express the Patched protein (see Table 2.2), a receptor for the Hh signal. Transduction of Hh signaling induces a stripe of cells on the anterior side of the A/P boundary to express the Dpp signaling protein (see Fig. 3.13a,b). Dpp, in turn, acts as a morphogen and diffuses across both the anterior and posterior compartments from its source along the A/P compartment boundary, thereby regulating gene expression in a concentration-dependent manner.

The dorsoventral coordinate system. The D/V coordinate system includes the sequential organizing activities of Apterous and the Notch and Wingless pathways.

The organization of the D/V axis is regulated in a fashion somewhat analogous to the regulation of the A/P axis organization. Dorsal cells are segregated from ventral cells by a lineage restriction imposed by the *apterous* compartment selector gene (see Table 2.1), which is expressed in all dorsal cells (see Fig. 13a). The Apterous protein regulates the expression of Serrate and Fringe (see Fig. 3.13b and Table 2.2), two proteins that interact with the ubiquitously expressed Notch receptor protein. Along both sides of the D/V boundary, Notch-mediated signaling between cells induces expression of the Wg signaling protein in a stripe that straddles the D/V boundary. Wg acts as a morphogen, diffusing from its source to regulate gene expression in both the dorsal and ventral compartments.

Signal integration by the vestigial field-specific selector gene. One important downstream target of the Dpp, N, and Wg organizing signals emanating from the compartment

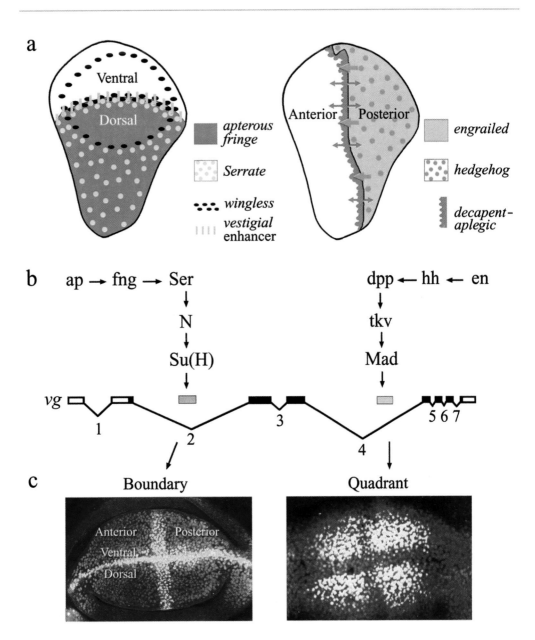

Figure 3.13
The genetic regulatory hierarchy in the *Drosophila* wing

(a) Two systems organize the pattern along the dorsoventral (**left**) and anteroposterior (**right**) axes of the wing. (b) Compartmental selector genes establish signaling sources along each compartment boundary. Transduction of these signals leads to the activation of downstream target genes. (c) One important target that is directly regulated by these signals is the *vestigial* wing selector gene. The *vg* gene is expressed in all wing cells through the sum of the activities of two *cis*-regulatory elements—one active along the compartment boundaries, and one active in the remaining four quadrants of the wing field.

boundaries is the *vg* field-specific selector gene. The *vg* gene is expressed in all of the cells in the wing disc that will form the wing. The regulation of the overall pattern of its expression is the sum of inputs from two separate *cis*-regulatory elements that respond to signals from each coordinate system (see Fig. 3.13c). The "boundary" *cis*-element is directly activated along the D/V boundary through the Notch pathway. The "quadrant" *cis*-element is expressed in the complementary pattern in the rest of the wing field through activation by the Dpp pathway and other regulators (see Fig. 3.13c). Expression of this field-specific selector gene depends directly on the signaling sources of the wing field and indirectly on the compartmental selectors that establish the expression of these signals.

Combinatorial regulation of wing patterning by signaling proteins and the Vg/Sd selector proteins

Within the wing field, a large number of genes function to pattern various features of the wing. For example, the AS-C genes are activated along the anterior extent of the D/V boundary to generate the sensory organs of the leading edge of the adult wing. The *blistered* or *Drosophila Serum Response Factor* (*D-SRF*) gene is expressed in all of the cells that lie between the developing wing veins. Many genes are involved in the positioning and differentiation of the veins that provide the structural support for and the fluid transport system of the wing. The development of each wing pattern element and the expression of the genes required for the elements' development depends on combinatorial inputs from signals and selector genes. In particular, the signals regulate where within the wing field a given gene is activated or repressed, and the activities of Vg and Scalloped (the DNA binding partner of Vg) make the response to these signals specific to the wing field.

The response to organizing signals. Various ligands for the major signaling pathways (Serrate, Dpp, Wnt, Hedgehog, and so on) are deployed in dynamic patterns in the wing field. Some signals appear to have greater range of influence than others do. The Hedgehog protein, for example, induces Dpp expression over a few cell diameters, whereas the Dpp signal produced within these cells influences cells as many as 20 or more cell diameters away.

At the level of gene regulation, the response to signaling inputs may be graded or threshold. In addition, some genes are activated at high concentrations of signal, whereas others are activated at a wider range of concentrations. The differential response of genes can create nested patterns of gene expression along a given axis of the wing field. For example, the *spalt* gene is activated at high Dpp levels, the *omb* gene at lower Dpp levels, and the *vg* quadrant element at still lower levels; consequently, the three genes are expressed in nested, partially overlapping domains centered on the A/P axis (Fig. 3.14). The different responses of target genes to levels of the Dpp signal fit the classic description of a morphogen.

The action of the Vg/Sd selector genes. Because the various signaling pathways deployed in the wing field are active elsewhere in the body, regulatory mechanisms must exist to impart specificity to their activities in any given tissue. This role is fulfilled by the field-specific selector genes. In the wing, the Vg and Sd proteins are required to permit appendage formation and to effect a wing-specific response of target genes to particular signals. The molecular basis for the Vg/Sd selector function is as follows: Sd acts as a sequence-specific DNA binding protein that binds to *cis*-regulatory elements that control

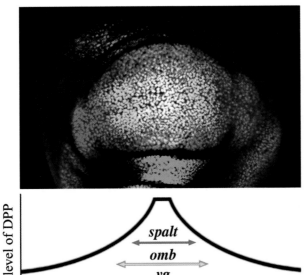

Figure 3.14
Activation of target genes by the Dpp morphogen gradient

The Dpp protein is expressed along the anterior/posterior compartment boundary of the wing, and spreads from this source to establish a concentration gradient. The *spalt* (**purple**), *opto-motor-blind* (*omb*; **yellow**), and *vg* (**red**) target genes are activated in nested patterns centered along this boundary. The differential response of target genes to a common signal is critical to pattern formation in cellular fields.

Source: Courtesy of Jaeseob Kim.

patterns of target gene expression in the wing. Sd forms a complex with Vg; in this complex, the Vg protein acts as a transcriptional activator. Importantly, however, the binding of Vg/Sd is not sufficient to activate target genes. A second cue is required from one or more signaling pathways or another regulatory input. This combinatorial regulation by the selector and signaling input restricts gene activity to particular cells within the wing field (Fig. 3.15).

Many genes that are expressed in discrete patterns in the wing field are also expressed in entirely different patterns in other fields. The wing-specific aspects of these genes' expression are controlled by independent wing-specific *cis*-regulatory elements that respond to and contain binding sites for Vg/Sd and at least one other input. Other *cis*-regulatory elements control gene expression elsewhere in the body. The modularity of the various *cis*-regulatory elements allows for the independent regulation of genes in the wing field and in other regions of the body.

The modification of regulatory hierarchies in secondary fields by Hox *genes*

The regulatory events in the growing wing field described previously transpire in the absence of *Hox* gene activity. Similarly, the patterning of the antennal field does not involve any *Hox* gene. On the other hand, the development of the haltere (the serial homolog of the wing) and the serial homologs of the antenna (which include all limb-type

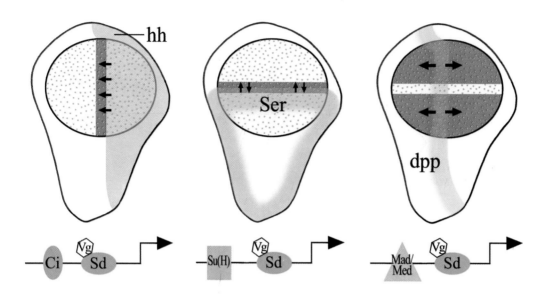

Figure 3.15
Integration of signaling and selector protein inputs by *cis*-regulatory elements

Target genes are activated in the wing field by a combination of inputs from one or more signal transducers and the Vg/Sd selector proteins. Localized activation of the Hedgehog (**left**), Notch (**middle**), or Dpp (**right**) signaling pathways induces target gene expression (**red**) where the signaling input and selector input (Vg/Sd; **blue stippling**) overlap. Each target gene possesses wing-specific *cis*-regulatory elements that integrate inputs from the appropriate signal transducer (Ci, Mad/Medea, Su(H); see Table 2.2) and Vg/Sd.

appendages) does involve *Hox* genes. The *Hox* genes aid the differentiation of these serial homologs by modifying developmental programs between serially homologous fields.

Before we discuss how *Hox* genes modify these complex morphologies and regulatory hierarchies, it is useful to briefly summarize our understanding of Hox protein functions. All *Hox* genes encode homeodomain proteins with similar DNA-binding specificities. One way that Hox proteins exert specific regulatory control is by interacting with protein co-factors. The Extradenticle (Exd) protein, another homeodomain-containing protein, is a key co-factor that interacts with several Hox proteins. Certain Hox-regulated *cis*-regulatory elements contain both Hox and adjacent Exd protein binding sites; occupancy of both sites and interaction between Hox and Exd molecules are required for gene regulation.

Hox proteins can act as both transcriptional activators and repressors, with their precise roles depending on a number of variables. In addition to co-factor interactions, post-translational modifications (such as phosphorylation) and the context of Hox binding sites can influence Hox activity. The core recognition sequence for Hox proteins is only about 6 bp long, so the genome contains many low- and high-affinity sites for these proteins.

The striking feature of *Hox* genes is that their expression in a new position can be sufficient to transform one structure into another. For example, expression of *Antp* in the antenna, where it is normally absent, causes the development of a leg. Similarly, expression of *Ubx* in the wing transforms that structure into a haltere. How can changes in a single gene so radically alter development? Do Hox proteins act globally to modify the expression of genes at the top of regulatory hierarchies? Or do they act throughout hierarchies, modifying some regulatory interactions but not others?

The picture that has emerged is one depicting *Hox* genes as "micromanagers" that act at many levels of regulatory hierarchies on selected components. The development of the haltere, for example, depends on the same major regulators (*apterous, engrailed, hedgehog, vestigial,* and so on) as its serial homolog, the wing. Yet, the morphology of the haltere is dramatically different in terms of its smaller size, balloon shape, lack of veins or large bristles, and details of cell architecture. The Ubx protein modifies the wing-patterning hierarchy to shape the development of the haltere by acting on a selected subset of genes that influence features of wing pattern formation (Fig. 3.16). For example, Ubx suppresses the production of the Wg protein in the posterior compartment of the haltere disc. This protein also represses AS-C genes, resulting in the suppression of bristle formation at the edge of the haltere. Along the A/P axis of the haltere, certain Dpp-regulated genes are specifically repressed by Ubx, while others are expressed in patterns similar to those in the wing.

How is the selective regulation of genes controlled by Ubx in the haltere? Once again, the key is *cis*-regulatory elements. Recall that wing-specific patterns of gene expression are controlled by discrete elements. In genes that are directly controlled by Ubx in the haltere, wing-specific *cis*-regulatory elements also contain sites for the Ubx protein that allow the expression of the gene to be selectively modified in the haltere. *The modularity of* cis-*regulatory elements enables the selective and differential regulation of gene expression by* Hox *proteins between serially homologous fields.*

THE VERTEBRATE BODY PLAN

The development of the vertebrate body plan has long been a focus of experimental embryology. Many fundamental concepts such as organizers, fields, and morphogens were derived first from observations of vertebrate embryos. The major features of adult vertebrate morphology, including segmented vertebral columns, paired appendages, and skulls, have undergone considerable evolutionary diversification. Therefore, we will focus on the developmental genetics of these major features here, and consider their evolutionary origins and modification in subsequent chapters.

Most vertebrate embryology has focused on a few amphibian (for example, *Xenopus* and newts), avian (for example, chick), mammalian (primarily the house mouse), and, more recently, fish (for example, zebrafish) species. The early development of these embryos and the mechanisms that orient the primary axes differ considerably, particularly between species with yolky eggs (for example, amphibians, fish) and species that rely upon extraembryonic structures. Nevertheless, all early vertebrate development converges on an embryonic stage in which the head is distinct, a neural tube extends along the dorsal midline above the notochord, and part of the mesoderm is subdivided. At this stage, the major anteroposterior axis (often called the "rostrocaudal" axis) is well defined, the general rostrocaudal domains of major head, trunk, and tail regulatory genes are initially determined, and the secondary fields (for example, limb buds) that will give rise to appendages and other structures are set to emerge. In the following discussion, we will see that the developmental regulatory logic and mechanisms controlling the patterning of vertebrate body axes and cellular fields revisit some already familiar themes mentioned earlier in this chapter.

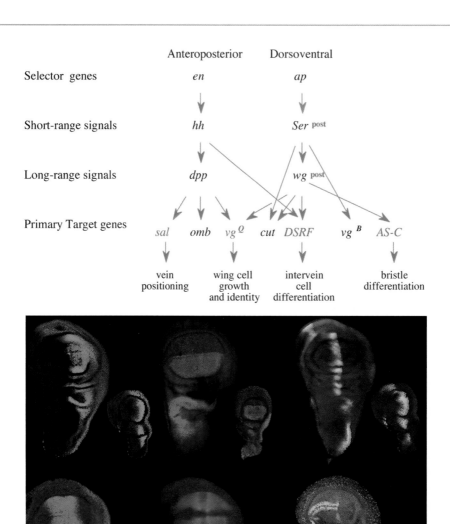

Anteroposterior   Dorsoventral

Selector genes   *en*   *ap*

Short-range signals   *hh*   *Ser* post

Long-range signals   *dpp*   *wg* post

Primary Target genes   *sal* *omb* *vg* ^{Q} *cut* *DSRF* *vg* ^{B} *AS-C*

vein   wing cell   intervein   bristle
positioning   growth   cell   differentiation
and identity   differentiation

Figure 3.16
The Ultrabithorax-regulated hierarchy in the *Drosophila* haltere

The haltere is a serial homolog of the wing. The Ubx protein regulates the morphological differentiation of the haltere by selectively modifying the wing regulatory hierarchy at many levels. (**top**) Genes or regulatory elements that are Ubx-regulated in the haltere are shown in red. (**bottom**) The expression patterns of genes in the wing and haltere imaginal discs. (**top row, left to right**) *engrailed*, *apterous*, and *dpp* expression are similar in the two fields. (**bottom row, left to right**) *spalt*, *vg* quadrant enhancer, and *achaete* expression differ between the wing and haltere discs due to the actions of Ubx.

Source: Modified from Weatherbee SD, Halder G, Hudson A, et al. Genes Develop 1998;12:1474–1482.

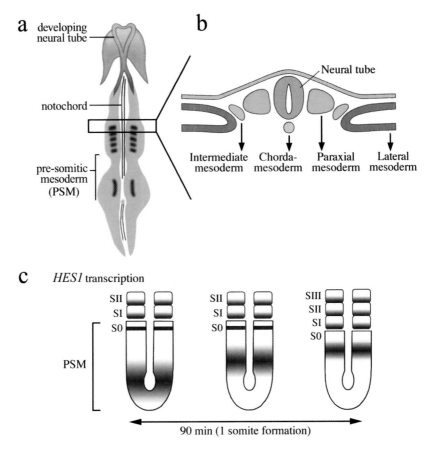

Figure 3.17
Segmentation and somite formation in the vertebrate embryo

(a) The segmentation of the paraxial mesoderm occurs in an anterior to posterior order. (b) The mesoderm of the early vertebrate embryo shown in cross section. (c) The *HES1* gene of the mouse is expressed in a dynamic pattern that cycles in the presomitic mesoderm once during the generation of each pair of somites.

Source: Part c adapted from Jouve C, Palmeirim I, Henrique D, et al. Development 2000;127:1421–1429.

The anteroposterior axis: somite formation and segmentation

The mesoderm of vertebrates gives rise to many organs and to the tissues that form much of vertebrate body architecture. The mesoderm is subdivided into several regions that give rise to different tissues (Fig. 3.17a,b). The **lateral plate mesoderm** is the source of elements of the circulatory system, the lining of body cavities, and all of the mesodermal components of the limbs (except the muscles). The **paraxial mesoderm** is subdivided into metameric subunits called **somites**. The first segmented structures to form, the somites provide the framework for the metameric organization of somite-derived tissues, including the axial skeleton (vertebrae), the dermis of the back, muscles, and the spinal ganglia. Because somite-derived structures such as the vertebrae have characteristic shapes in different regions of the anteroposterior axis, the generation and individuation of the somites and their derivatives are fundamental to the evolution and diversification of the vertebrate body plan.

Segmentation of the paraxial mesoderm and somite formation represent a sequential, temporally ordered process that proceeds from the anterior to the posterior of the embryo. New somites are formed from the rostral end of the paraxial mesoderm at regular intervals. In the mouse and the chick, a defined, albeit different number of somites are produced, at roughly 90-minute intervals.

Two regulatory systems have been identified that underlie the temporal and spatial regulation of somite formation. One system involves genes that are expressed in oscillating patterns during the progression of somite formation. The second system behaves as a segmentation clock, controlling the oscillating expression patterns of these genes in the developing somites.

The *HES1* gene is one of several genes whose expression oscillates during somite formation in the mouse. A relative of the *Drosophila hairy* gene, *HES1* is expressed in a dynamic wave that sweeps across the presomitic mesoderm once during the formation of each somite (Fig. 3.17c). *HES1* expression is regulated by components of the Notch pathway in the mouse. Indeed, oscillating gene expression during somite formation in all vertebrates appears to be regulated by the Notch pathway.

The vertebrate Hox *ground plan*

The somites of the vertebrate embryo give rise to the major axial structures, including the vertebrae, ribs, and skeletal muscles, as well as the dermis. The vertebrae of the spinal column and their associated processes are of five distinct types: cervical, thoracic, lumbar, sacral, and caudal (Fig. 3.18a). The transitions between these types occur at specific somite positions in each species. Similarly, the forelimb and hindlimb buds, which develop from the unsegmented lateral plate mesoderm, arise at specific axial positions. Therefore, the morphologies of somite derivatives at different axial levels must be genetically regulated.

Patterning along the rostrocaudal axis in all vertebrates is regulated by the *Hox* genes. Vertebrates have four or more complexes of *Hox* genes; the mouse, for example, has 39 *Hox* genes. In general, the anterior boundaries of expression of the *Hox* genes correlate with the respective locations of the genes within a complex, a principle termed *colinearity*. Thus the *Hox1* genes (which are related to the *labial* gene) are expressed in the most anterior positions, and the *Hox9-13* genes (which are related to the *Abd-B* gene) are expressed in more posterior positions (see Fig. 2.22b). The anterior boundaries of *Hox* gene expression are generally sharper than their posterior boundaries and lie in spatial register between the somitic and lateral plate mesoderm. However, they are not in register with *Hox* gene expression in the neural tube. The tissue-specific boundaries of *Hox* gene expression indicate that independent genetic regulatory mechanisms control *Hox* expression in the neural tube.

Patterning of part of the neural tube, the posterior region of the head, and the hindbrain also involves segmentation along the A/P axis. The process of hindbrain segmentation is evolutionarily conserved in vertebrates in terms of the number of subdivisions or **rhombomeres** formed (seven to eight), the neuroanatomical organization of each region, and the pattern of expression of various genes, including transcription factors and signaling proteins that establish rhombomere segmentation, identity, and cell behavior. The best-known regula-

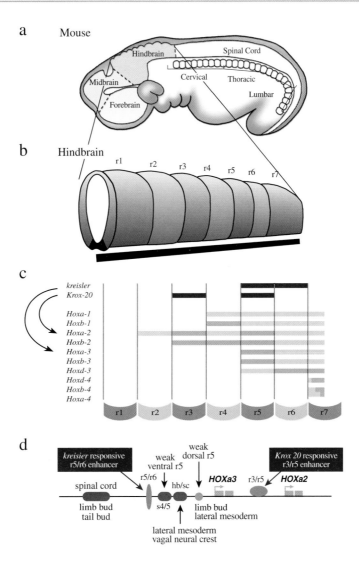

Figure 3.18
Regulation of *Hox* gene expression in the vertebrate hindbrain

(**a**) The vertebrate hindbrain forms anterior to the spinal cord. (**b**) The hindbrain is overtly segmented into rhombomeres (r1–r7). (**c**) Summary of regulatory gene expression in r1–r7. The arrows denote regulatory interactions between two transcription factors (Kreisler and Krox20) and two *Hox* genes of interest. (**d**) A schematic of the genetic region around the *Hoxa2* and *Hoxa3* genes. Specific *cis*-regulatory elements (R3 and R5) control the response of *Hoxa2* to Krox20 in r3/r5. Note the presence of several other elements controlling *Hox* expression in other tissues. This diversity of *cis*-elements is typical of *Hox* genes.

Source: Modified from Lumsden A, Krumlauf R. Science 1996;274:1109–1115; Manzanares M, Cordes S, Ariza-McNaughton L, et al. Development 1999;126:759–769; Nonchev S, Vesque C, Maconochie M, et al. Development 1996;122:543–554.

tory genes involved in this process are the *Hox1* through *Hox4* genes, various members of which are expressed in and affect the development of discrete rhombomeres (see Fig. 3.18a,b).

The same themes pertain to *Hox* regulation in the vertebrate hindbrain (and elsewhere) as were highlighted for the *Drosophila Hox* genes:

- Numerous independent *cis*-acting regulatory elements control features of individual *Hox* gene expressions in different rostrocaudal domains and in different germ layers.

- Both positive and negative regulatory interactions are necessary to define the initial domains of gene expression.

- Autoregulatory and cross-regulatory interactions between *Hox* genes are necessary to maintain domains of gene expression.

In the developing mouse hindbrain, a few direct transcriptional regulators of *Hox* gene expression patterns have been identified, including the products of the *Krox20* and *Kreisler* genes, and products of *Hox* genes themselves. In addition, *cis*-regulatory elements have been identified that control rhombomere-specific patterns of *Hox* gene expression (Fig. 3.18d). For example, expression of the *Hoxa2* and *Hoxb2* paralogs in rhombomeres 3(r3) and 5(r5) are controlled by distinct elements, independent of those that control expression in the mesoderm and other tissues. The Krox20 protein is specifically expressed in these two rhombomeres and directly regulates the r3/r5 enhancers (Fig. 3.18c). Similarly, expression of the *Hoxa3* and *Hoxb3* genes in r5/r6 is controlled by discrete elements. The Kreisler protein is specifically expressed in these developing rhombomeres, and it binds to and directly regulates *Hoxa3* and *Hoxb3* expression through these elements (see Fig. 3.18c).

Apart from these rhombomere patterns, not much is known about the transcription factors that position *Hox* domains along the rostrocaudal axis. From the *cis*-regulatory elements of the *Hox* complexes, however, we can infer that the regulatory network likely involves many interactions. When the intergenic regions of a few pairs of *Hox* genes were analyzed in greater detail, a substantial number of regulatory elements were found to be interspersed among the *Hox* genes. Separate *cis*-regulatory elements that control gene expression in the lateral mesoderm, neural crest, spinal cord, tail bud, limb buds, and somites have been identified, for instance (see Fig. 3.18d). The temporal, spatial, and tissue-specific aspects of each pattern must be regulated by a variety of transcription factors acting through these elements. The different anterior expression boundaries of adjacent genes within complexes could reflect the differential response of *cis*-regulatory elements to the same regulator or the specific response of each element to different regulators. Interestingly, some of the *cis*-acting regulatory elements are shared between pairs of adjacent genes, and this sharing may be a mechanism that preserves the clustering of *Hox* genes.

Vertebrate limb development

Among the most important structures in vertebrate evolutionary history are the paired pelvic and pectoral appendages. The evolution of the tetrapod limb from the paired fins of fish and the subsequent modification of limb morphologies used in flying, running and jumping, burrowing, and the return to water involved major changes in the anatomy of paired appendages. Coupled with the long-standing research on the experimental embryology of limb buds, immense interest is focused on the developmental genetics of vertebrate limb formation and patterning. For our purposes, there are six major aspects of limb formation and patterning, each of which represents a potential point of divergence between taxa in the functional evolution of limb morphology:

- The positioning of the limb bud along the rostrocaudal axis
- The initial outgrowth of the limb bud
- Establishment and signaling from the limb organizers
- The formation of the limb proximodistal axis and deployment of *Hox* gene expression in the limb bud

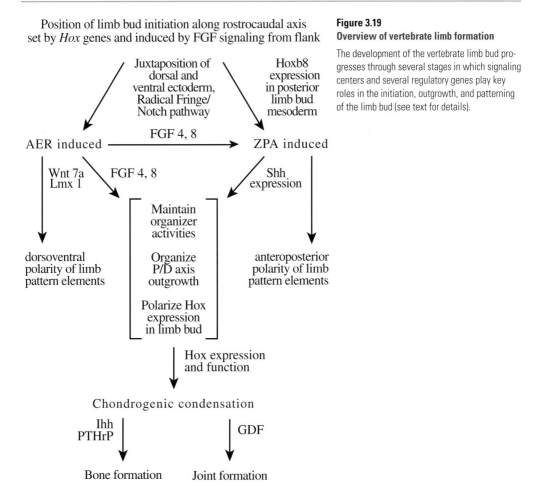

Figure 3.19
Overview of vertebrate limb formation

The development of the vertebrate limb bud progresses through several stages in which signaling centers and several regulatory genes play key roles in the initiation, outgrowth, and patterning of the limb bud (see text for details).

- The specification of major skeletal and cartilage elements
- The regulation of forelimb and hindlimb identity

These developmental processes unfold as a cascade of events that progressively establish the limb field, its two organizers, and many pattern elements (Fig. 3.19).

The developmental regulatory logic involved in these processes of limb formation and patterning are analogous to those discussed for insects. The position of the limb field depends on signals whose distribution is regulated along the main body axis. The outgrowth of the limb field involves regulatory hierarchies that establish the anteroposterior and dorsoventral axes of the limb field, and integration of these inputs regulates patterning along the proximodistal axis. Superimposed on the potentially equivalent hierarchies in the forelimb and hindlimb fields are selector genes that specifically modify the development of these limbs.

Positioning of the limb buds along the rostrocaudal axis

Limb development begins with the proliferation of mesenchymal cells from within the limb field of the lateral plate mesoderm. As these cells accumulate under the epidermis, a bulge appears—the growing limb bud. This proliferation appears to be under the control of local signals emanating from nearby mesoderm. One strong candidate for the source of this signal is fibroblast growth factor 8 (FGF8), which is expressed near the site of limb bud initiation and can induce the formation of ectopic limb buds when expressed at novel positions along the rostrocaudal axis (Fig. 3.20a).

Limbs of different vertebrates arise at different positions, with respect to somite number, along the rostrocaudal axis. Nevertheless, the forelimb buds always arise at the most anterior position of *Hoxc6* expression at the transition from the cervical vertebra to the thoracic vertebra. This consistent location suggests that the positioning of the limb bud is determined by both axial coordinates (*Hox* genes) and local expression of FGF8. Indeed, regulation of FGF8 expression by the axial *Hox* genes in the mesoderm may dictate where limb buds form.

Outgrowth of the limb bud: dorsoventral and anteroposterior regulatory hierarchies

The early limb bud consists of mesodermal cells and an overlying ectodermal epithelium. As it forms, the mesodermal cells induce the overlying ectoderm to form the apical ectodermal ridge (AER), a prominent thickening of cells at the edge of the limb bud where its dorsal and ventral halves meet. The AER organizer is a major signaling center that produces signals such as FGF8 (and other related FGFs), which are necessary for the continued proliferation and differentiation of the growing limb bud (Fig. 3.20b). Loss of the AER, or its signaling functions, results in truncation of the limb along the proximodistal axis.

The induction of the AER depends on the juxtaposition of dorsal and ventral ectoderm (Fig. 3.21). The expression of the signaling protein Radical Fringe, a molecule that interacts with the vertebrate form of the Notch protein, is restricted to the dorsal ectoderm; it mediates the selective induction of FGF8 expression and AER formation at the D/V boundary of the limb ectoderm. The restriction of *Radical fringe* expression to the dorsal ectoderm is regulated by En-1, a homolog of the *Drosophila engrailed* gene, which is expressed in all ventral ectodermal cells (see Fig. 3.21b).

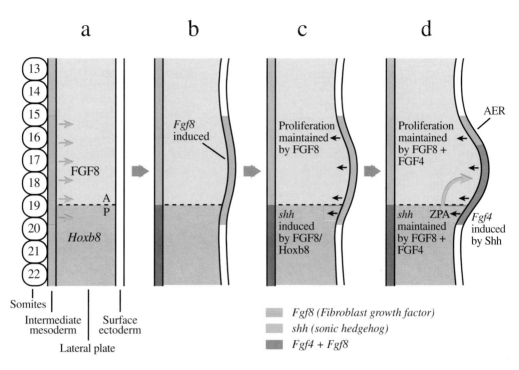

Figure 3.20
Regulation of vertebrate limb bud formation

The vertebrate limb bud is induced by the localized expression of FGF8 in the mesoderm. FGF8 is induced in the overlying surface ectoderm in the AER and maintains proliferation of the bud. Shh is induced in the posterior of the limb bud by FGF8 and Hoxb8. The two signaling sources are maintained by reciprocal interactions.

Source: Modified from Gilbert S. Developmental biology, 5th ed. Sunderland: Sinauer Associates, 1997.

En-1 also regulates dorsal patterning of the limb, which occurs later during limb bud outgrowth. Different pattern elements are formed by dorsal and ventral regions of the limb bud. En-1 represses expression of the Wnt-7a protein (a member of the signaling protein family to which the insect Wg protein belongs), which is then expressed only in the dorsal ectoderm and regulates dorsoventral polarity. The actions of Wnt-7a appear to be mediated by the Lmx-1 protein, a homolog of the *Drosophila apterous* gene that is expressed in all dorsal mesodermal cells (see Fig. 3.21b).

Patterning along the A/P axis of the limb bud depends on the function of the zone of polarizing activity (ZPA), another organizer with potent signaling activity. The ZPA is the source of the Sonic hedgehog (Shh) protein in the limb field. The ZPA and Shh influence the polarity of pattern elements along the A/P axis. In particular, digits with reversed polarity can be induced when either the ZPA is transplanted or Shh expression is induced in the anterior part of the limb.

a dorsal

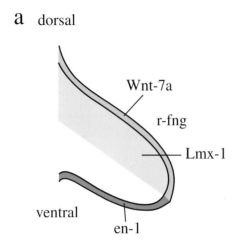

Wnt-7a

r-fng

Lmx-1

ventral

en-1

b

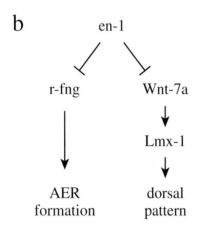

en-1

r-fng   Wnt-7a

Lmx-1

AER   dorsal
formation   pattern

Figure 3.21
AER formation and dorsoventral patterning of the limb bud

(a) Gene expression differs along the D/V axis of the limb bud. (b) The regulation of AER formation and dorsal patterning depends on the action of the *en-1* gene in the ventral ectoderm. The *en-1* gene product prevents *r-Fng* and *Wnt-7a* expression, which function to organize AER formation and dorsal patterning, respectively.

Source: Johnson RL, Tabin CJ. Cell 1997;90: 979–990.

The restriction of the ZPA and Shh expression to the posterior portion of the limb bud suggests that the limb bud contains anteroposterior positional information that precedes Shh expression and function. This preexisting information may consist of rostrocaudal regulatory patterns derived from the lateral plate mesoderm. Indeed, the anterior limit of *Hoxb8* gene expression in the lateral plate coincides precisely with the anterior limits of Shh expression in the early limb bud. Furthermore, ectopic expression of *Hoxb8* in more anterior cells under the AER induces ectopic Shh expression. Therefore, Shh expression in the ZPA appears to be regulated both by signals from the AER and by Hox regulatory inputs in the lateral plate mesodermal cells that will become incorporated into the limb bud (see Fig. 3.20c).

The organizers in the limb field do not act independently of one another. Formation and maintenance of the ZPA depend on the presence of a functional AER, and, in turn, maintenance of the AER depends on the existence of ZPA function. Thus a regulatory circuit connects FGF signaling to the induction of Shh expression, and Shh signaling to FGF expression in the AER (see Fig. 3.20d). As yet, we do not know whether these inductive interactions occur directly or operate through intermediate signals.

Integration of organizing signals to form the proximodistal axis and deployment of Hox genes in the limb field

After the formation of the ZPA and AER and the proliferation of mesodermal cells, proximal mesenchymal cells begin to form cartilaginous condensations that prefigure the limb skeletal pattern. These condensations arise in a proximal to distal order, so that the humerus (the bone of the upper arm) forms first, the radius and ulna next, and the carpals (wrist bones) and digits last. The development of these individual pattern elements suggest that limb development goes through discrete temporal phases and that pattern formation is controlled by the localized expression of regulatory genes, cued by the signals emanating from the organizers, the AER and the ZPA.

The *Hox* genes play important roles in limb patterning, albeit in a different fashion than in appendage patterning in arthropods. Whereas individual *Hox* genes or pairs of *Hox* genes are expressed in arthropod limb fields, a much larger number of *Hox* genes are deployed in nested domains of vertebrate limb fields (Fig. 3.22). Detailed studies of normal and ectopic *Hox* gene expression and analysis of limb development in mice lacking one or more *Hox* gene functions have revealed a complex and dynamic spatiotemporal pattern of *Hox* gene patterns and gene interactions in determining the formation of limb pattern elements.

Three temporal phases of *Hox* gene expression appear to correlate with the temporal sequence of proximodistal pattern element formation, particularly in regard to the *Hox9-13* genes of the *HoxA* and *HoxD* complexes (see Fig. 3.22a–c). The first phase of *Hox* expression is not polarized, appears to be Shh-independent, and is associated with the development of the most proximal limb elements (upper arm/leg) (see Fig. 3.22a). Subsequent phases of *Hox* expression arise in nested patterns whose polarity depends on AER and ZPA functions. The second phase of *Hox* expression occurs in the next most proximal elements (forearm or lower leg) (see Fig. 3.22b). The third phase of *Hox* expression includes most distal elements (wrist and hand, ankle and foot) (see Fig. 3.22c).

These spatial patterns of *Hox* gene expression reveal that many related Hox proteins are expressed in overlapping as well as adjacent domains. No simple one-to-one correspondence exists between a particular *Hox* gene and the growth and patterning of any particular limb pattern element. Rather, the formation and identity of particular elements appear to reflect combinations of Hox protein functions, some of which are clearly acting redundantly. For example, loss of either *Hoxa-13* or *Hoxd-13* function has limited effects, whereas loss of both genes dramatically affects the development of the distal limb. Similarly, *Hoxd11* or *Hoxa11* mutations have minor effects on the formation of the radius and ulna, whereas these long bones are almost completely lost in the double mutant.

The analysis of *cis*-acting regulatory elements that control *Hox* gene expression in the

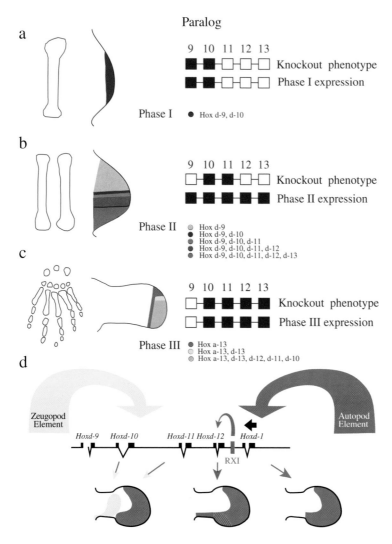

Figure 3.22

Hox **gene expression and function in vertebrate limb development**

(a–c) Three temporal phases of *Hox* gene expression in the limb bud correlate with the elaboration of three distinct elements of the limb. (a) Phase I involves a more limited set of *Hox* genes expressed across the entire bud in the region that will give rise to the upper arm or thigh (stylopod). Mutations in these genes primarily affect this structure. (b) Phase II expression is initiated in response to Sonic hedgehog signaling and is polarized with respect to the A/P axis of the limb bud. Disruption of multiple *Hox* genes that have pronounced phase II patterns affects formation of the bones corresponding to the lower arm or calf (zeugopod). (c) Phase III expression occurs in the hand and foot (autopod). Disruption of certain *Hox* genes primarily affects autopod formation and patterning. (d) The *cis*-regulatory elements of the *Hoxd9-13* genes. Both global and local elements regulate *Hox* expression in the limb bud. Multiple *Hox* genes are influenced by remote elements active in the zeugopod and autopod. Local elements, such as RXI (red), affect expression of individual genes in restricted domains.

Source: Shubin N, Tabin C, Carroll S. Nature 1997;388:639–648; Hérault Y, Beckers J, Kondo T, et al. Development 1998;125: 1669–1677.

limb bud has highlighted the complexity of *Hox* expression. Both gene-specific and global regulatory elements appear to control the expression of individual *Hox* genes and groups of *Hox* genes in the limb bud, respectively. For example, phase II expression of the *Hoxd12* gene is regulated by both a nearby element and a remote global element, whereas phase III expression is controlled by a remote global element (see Fig. 3.22d). Thus the complex patterns of *Hox* gene expression are built up from many potentially independent inputs, which creates the potential for morphological diversification through changes in the relative timing and spatial regulation of individual elements.

Regulatory networks controlling the differentiation of major limb pattern elements

The skeletal defects caused by *Hox* mutations indicate that one function of Hox proteins may be to regulate the position and timing of cartilage development and differentiation in the limb field. Evidence suggests that *Hox* genes, and another member of the Hedgehog signal family, Indian hedgehog (Ihh), regulate the progression of bone differentiation and thus major features of limb pattern formation. Certain Hox proteins appear to act early in the differentiation pathway to prevent its progression, either by promoting the proliferation of bone forming cells or by preventing their differentiation.

The positions of the joints between limb skeletal elements are also spatially regulated. One regulator of this process, the GDF5 signaling protein (a member of the large TGF-β superfamily), is specifically expressed in regions where bone development does not occur. GDF5 expression presages the location of joints throughout the limb field (as well as elsewhere in the body). Clearly, the regulation of the spatial expression of these various cartilage and bone promoting and inhibiting factors is key to the skeletal pattern formed.

The regulation of forelimb and hindlimb identity by selector genes

The serially homologous tetrapod forelimb and hindlimb are believed to have evolved from the paired pectoral and pelvic fins of fish, respectively. Although the developing forelimbs and hindlimbs both utilize organizing signals that are necessary for appendage outgrowth, no single transcription factor required for limb outgrowth has been identified as yet. The *Hox* genes, for example, do not play the role of selector genes for limb field formation or identity. No *Hox* mutations result in homeotic transformations between limb types.

A few other transcription factors have, however, been identified as selectors of forelimb and hindlimb identity in vertebrate limbs. The paired homeodomain-containing protein Pitx-1 and the T-box-containing protein Tbx4 are expressed specifically in the hindlimb mesenchyme, and the T-box gene *Tbx5* is expressed specifically in the forelimb mesenchyme (Fig. 3.23). The differentiation between chick forelimb (wing) and hindlimb (leg) identity appears to be regulated, at least in part, by these selector genes. Although each is required for proper identity, none of the three genes is sufficient by itself to confer limb-specific identity. Hence, additional regulators of limb identity may act upstream of these genes, or perhaps limb identity results from a number of field-specific inputs.

These and other potential field-specific regulators presumably function to control the differential expression of genes between the two fields (see Fig. 3.23). Several *Hox* genes, including *Hoxc10* and *Hoxc11*, are differentially expressed between limb types and are potential targets of the limb-specific selectors. Given the important morphological and

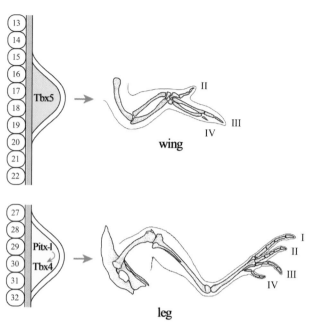

Figure 3.23
Selector genes controlling vertebrate limb identity

The differential development of the vertebrate forelimb and hindlimb is under the control of selector genes that are differentially expressed in the limb mesenchyme. The *Tbx5* gene is expressed in the forelimb; the *Pitx-1* and *Tbx-4* genes are expressed in the developing hindlimb. In the chick, products of these genes regulate wing versus leg identity.

functional diversity of vertebrate limb morphologies, it will be important to characterize the levels in the limb regulatory hierarchy at which changes have arisen between taxa.

REVIEW: THE GENERAL LOGIC AND MECHANISMS CONTROLLING GENE EXPRESSION IN CELLULAR FIELDS

This chapter has presented an overview of the genetic regulatory logic and mechanisms that operate to control development in a few model species. It concentrated on the genes of the general animal toolkit for development as well as the mechanisms that control large-scale patterning of the main body axes and secondary fields. Although we explored only a modest number of regulatory hierarchies in representatives of just two phyla, the similarities in the regulatory logic and mechanisms allowed us to identify some general themes concerning the regulation and function of toolkit genes and the architecture of the regulatory hierarchies that progressively specify pattern formation in cellular fields. The identification of these general themes is critical for understanding trends in animal evolution.

Because the regulatory mechanisms of development are themselves the product of evolution, their architecture is a reflection of the evolutionary processes that assembled them. Armed with our knowledge of regulatory architectures in one species, we can begin to make comparisons with other taxa with different morphologies to identify at what level and through what genetic mechanisms developmental and morphological diversity arises.

From this survey of regulatory hierarchies and mechanisms, we underscore three general themes with regard to the architecture of regulatory hierarchies, the molecular mechanisms controlling gene expression, and the functions of toolkit genes in controlling gene expression and patterning in cellular fields:

1. *The development of the growing embryo and its body parts occurs in a spatially and temporally ordered progression through the sequential generation of coordinate systems.*

    The position of any adult pattern element is determined through a series of hierarchies that subdivide the embryo and its morphogenetic fields into progressively finer elements. From the coordinate systems that subdivide the major embryonic axes, secondary fields are specified that contain their own autonomous coordinate systems. Domains within these fields can be further subdivided and specified through the establishment of local domains of gene expression. Ultimately, the temporal and spatial segregation of these processes allows for the modular organization of bilaterians and the individualization of body parts and pattern elements, such that morphology evolves independently from the rest of a field or the body plan.

2. *The modularity of* cis-*regulatory elements acting on individual genes allows for the independent spatial and temporal control of discrete features of gene expression and function.*

    Many gene expression patterns are actually the sum of the functions of many independent *cis*-regulatory elements. The modularity of *cis*-regulation is crucial to the control of the specificity of gene interactions and function during development. The independent spatial and/or temporal regulation of gene expression permits individual genes to have different, but specific functions in different contexts. Thus, while operating through identical signal transduction pathways, a signaling protein can be expressed in entirely unrelated populations of cells in two different tissues (by virtue of its own *cis*-elements) and can regulate the expression of completely different target genes (through different *cis*-regulatory elements of its target genes).

    In this light, it is not adequate or accurate to describe a given toolkit gene function solely in terms of the protein it encodes, because the function of that protein almost always depends on the context in which it is expressed. Instead, toolkit genes should be viewed as consisting of both a functional protein and a suite of regulatory elements that control its deployment. Each element represents a separate genetic function.

    The modular organization of gene-specific control elements has two profound evolutionary implications. First, this modularity is the product of evolution and therefore directly reflects the mechanisms that allow new and independent patterns of gene expression to evolve. Second, the modular control of extant genes allows for changes in one aspect of gene expression and function without affecting any other functions. In other words, modularity facilitates the dissociation of gene functions as well as the evolution of new interactions and potential morphologies.

3. *Spatial patterns are the product of combinatorial regulation.*

    In the examples of regulatory hierarchies and individual genes expression patterns analyzed in this chapter, we saw that new patterns were created by the combined inputs of preceding patterns. The integration of D/V and A/P inputs to position primordia, the positioning of a pair-rule stripe, the positioning of organizers, and the field-specific pattern of a regulatory gene, for example, are all derived from the integration of multiple inputs (some positive, some negative) by *cis*-regulatory elements.

Combinatorial control is key to both the *specificity* and the *diversity* of gene expression. In terms of specificity, it provides a means to localize gene expression to discrete cell populations using inputs (such as signaling pathways) that are not cell-type-specific or tissue-specific. In terms of diversity, combinatorial mechanisms provide a means to generate a virtually limitless variety of spatial patterns through the overlapping inputs of positive and negative regulators and autoregulatory feedback mechanisms. As the number of discrete domains of gene activity increases in a field, the potential combinations of regulators and patterning outputs increase exponentially.

The evolutionary significance of combinatorial regulation is obvious. New gene expression patterns can evolve as new combinations of regulatory inputs are integrated. In turn, the new gene expression patterns create the potential for further change, and the cycle continues.

From these general themes, we can start to anticipate how regulatory hierarchies and gene expression patterns might be cobbled together in the course of evolution and how they might change. To begin to understand what has occurred, we must take a broad inventory of the genetic toolkit across the animal kingdom and examine the regulatory mechanisms for building animals in a phylogenetic and comparative context to see how morphological diversity evolves.

SELECTED READINGS

Gene Regulation in Metazoans

Arnone, M. I., Davidson, E. H. The hardwiring of development: organization and function of genomic regulatory systems. *Development* 1997; 124:1851–1864.

Gilbert, S. F. Developmental biology, 5th ed. Sunderland: Sinauer Associates, 1997.

Kornberg, R. D. Eukaryotic transcriptional control. *TCB-TIBS-TIG,* millennium issue 2000:M46–M49.

Ptashne, M. A genetic switch: phage λ and higher organisms. Cambridge, MA: Blackwell Scientific, 1992.

Wolpert, L., Beddington, R., Brockes, J., et al. Principles of development. London: Current Biology Ltd., 1998.

From Egg to Segments: The Anteroposterior Coordinate System

Burz, D. S., Rivera-Pomar, R., Jäckle, H., Hanes, S. D. Cooperative DNA-binding by Bicoid provides a mechanism for threshold-dependent gene activation in the *Drosophila* embryo. *EMBO J* 1998; 17:5998–6009.

Driever, W. In: Bate, M., Arias, A. M., eds. The development of *Drosophila melanogaster,* vol. 1. Cold Spring Harbor: Cold Spring Harbor Laboratory Press, 1993:301–324.

Driever, W., Nüsslein-Volhard, C. The bicoid protein is a positive regulator of *hunchback* transcription in the *Drosophila* embryo. *Nature* 1989; 337:138–143.

Rivera-Pomar, R., Jäckle, H. From gradients to stripes in *Drosophila* embryogenesis: filling in the gaps. *Trends Genetics* 1996; 12:478–483.

Small, S., Blair, A., Levine M. Regulation of *even-skipped* stripe 2 in the *Drosophila* embryo. *EMBO J* 1992; 11:4047–4057.

Stanojevic, D., Small, S., Levine, M. Regulation of a segmentation stripe by overlapping activators and repressors in the *Drosophila* embryo. *Science* 1991; 254:1385–1387.

The Dorsoventral Axis Coordinate System

Jiang, J., Levine, M. Binding affinities and cooperative interactions with bHLH activators delimit threshold responses to the dorsal gradient morphogen. *Cell* 1993; 72:741–752.

Rusch, J., Levine, M. Threshold responses to the dorsal regulatory gradient and the subdivision of primary tissue territories in the *Drosophila* embryo. *Curr Op Genetics Develop* 1996; 6:416–423.

The Hox *Ground Plan*

Casares, F., Sánchez-Herrero, E. Regulation of the *infraabdominal* regions of the bithorax complex of *Drosophila* by gap genes. *Development* 1995; 121:1855–1866.

Müller, J., Bienz, M. Sharp anterior boundary of homeotic gene expression conferred by the *fushi tarazu* protein. *EMBO J* 1992; 11:3653–3661.

Qian, S., Capovilla, M., Pirrotta, V. Molecular mechanisms of pattern formation by the BRE enhancer of the *Ubx* gene. *EMBO J* 1993; 12:3865–3877.

Shimell, M. J., Simon, J., Bender, W., O'Connor, M. B. Enhancer point mutation results in a homeotic transformation in *Drosophila*. *Science* 1994; 264:968–971.

Zhang, C-C., Müller, J., Hoch, M., Jäckle, H. Target sequences for *hunchback* in control region conferring *Ultrabithorax* expression boundaries. *Development* 1991; 113:1171–1179.

Secondary Fields: Integrating the Anteroposterior and Dorsoventral Coordinate Systems

Cohen, B., Simcox, A. A., Cohen, S. M. Allocation of the thoracic imaginal primordia in the *Drosophila* embryo. *Development* 1993; 117:597–608.

Goto, S., Hayashi, S. Specification of the embryonic limb primordium by graded activity of Decapentaplegic. *Development* 1997; 124:125–132.

Kuo, Y. M., Jones, N., Zhou, B., et al. Salivary duct determination in *Drosophila*: roles of the EGF receptor signaling pathway and the transcription factors Fork head and Tracheales. *Development* 1996; 122:1909–1917.

Michelson, A. M. Muscle pattern diversification in *Drosophila* is determined by the autonomous function of homeotic genes in the embryonic mesoderm. *Development* 1994; 120:755–768.

Panzer, S., Weigel, D., Beckendorf, S. K. Organogenesis in *Drosophila melanogaster*: embryonic salivary gland determination is controlled by homeotic and dorsoventral patterning genes. *Development* 1992; 114:49–57.

Vachon, G., Cohen, B., Pfeifle, C., et al. Homeotic genes of the bithorax complex repress limb development in the abdomen of the *Drosophila* embryo through the target gene *Distal-less*. *Cell* 1992; 71:437–450.

Yagi, Y., Suzuki, T., Hayashi, S. Interaction between *Drosophila* EGF receptor and *vnd* determines three dorsoventral domains of the neuroectoderm. *Development* 1998; 125:3625–3633.

The Wing Field

Basler, K., Struhl, G. Compartment boundaries and the control of *Drosophila* limb pattern by hedgehog protein. *Nature* 1994; 368:208–214.

Blair, S. Compartments and appendage development in *Drosophila*. *BioEssays* 1995; 17:299–309.

Dahmann, C., Balser, K. Compartment boundaries at the edge of development. *Trends in Genetics* 1999; 15:320–326.

Halder, G., Polaczyk, P., Kraus, M. E., et al. The Vestigial and Scalloped proteins act together to directly regulate wing-specific gene expression in response to signaling proteins. *Genes Develop* 1998; 12:3900–3909.

Jouve, C., Palmeirim, I., Henrique, D., et al. Notch signalling is required for cyclic expression of the hairy-like gene *HES1* in the presomitic mesoderm. *Development* 2000; 127:1421–1429.

Kim, J., Johnson, K., Chen, H. J., et al. MAD binds to DNA and directly mediates activation of vestigial by DPP. *Nature* 1997; 388:304–308.

Lecuit, T., Brook, W., Ng, M., et al. Two distinct mechanisms for long-range patterning by Decapentaplegic in the *Drosophila* wing. Nature 1996; 381:387–393.

Nellen, D., Burke, R., Struhl, G., Basler, K. Direct and long-range actions of a Dpp morphogen gradient. *Cell* 1996; 85:357–368.

Neumann, C., Cohen, S. Morphogens and pattern formation. *BioEssays* 1997; 19:721–729.

Takke, C., Campos-Ortega, J. A. *her1*, a zebrafish pair-rule like gene, acts downstream of notch signalling to control somite development. *Development* 1999; 126:3005–3014.

Williams, J., Paddock, S., Vorwerk, K., Carroll, S. Organization of wing formation and induction of a wing-patterning gene at a compartment boundary. *Nature* 1994; 368:299–305.

Zecca, M., Basler, K., Struhl, G. Direct and long-range action of a wingless morphogen gradient. *Cell* 1996; 87:833–844.

———. Sequential organizing activities of engrailed, hedgehog and decapentaplegic in the *Drosophila* wing. *Development* 1995; 121:2265–2278.

Hox *Genes' Modification of Regulatory Hierarchies in Secondary Fields*

Akam, M. *Hox* genes: from master genes to micromanagers. *Curr Biol* 1998; 8:R676–R678.

Weatherbee, S. D., Carroll, S. B. Selector genes and limb identity in arthropods and vertebrates. *Cell* 1999; 97:283–286.

Weatherbee, S., Halder, G., Hudson, A., et al. Ultrabithorax regulates genes at several levels of the wing-patterning hierarchy to shape the development of the *Drosophila* haltere. *Genes Develop* 1998; 10:1474–1482.

The Vertebrate Hox *Ground Plan*

Lumsden, A., Krumlauf, R. Patterning the vertebrate neuraxis. *Science* 1996; 274:1109–1114.

Manzanares, M., Cordes, S., Ariza-McNaughton, L., et al. Conserved and distinct roles of *kreisler* in regulation of the paralogous *Hoxa3* and *Hoxb3* genes. *Development* 1999; 126:759–769.

Nonchev, S., Maconochie, M., Vesque, C., et al. The conserved role of *Krox-20* in directing *Hox* gene expression during vertebrate hindbrain segmentation. *Proc Natl Acad Sci USA* 1996; 93:9339–9345.

Pöpperl, H., Blenz, M., Studer, M., et al. Segmental expression of *Hoxb-1* is controlled by a highly conserved autoregulatory loop dependent upon *exd/pbx*. *Cell* 1995; 81:1031–1042.

Sharpe, J., Nonchev, S., Gould, A., et al. Selectivity, sharing and competitive interactions in the regulation of *Hoxb* genes. *EMBO J* 1998; 17:1788–1798.

Vesque, C., Maconochie, M., Nonchev, S., et al. *Hoxb-2* transcriptional activation in rhombomeres 3 and 5 requires an evolutionarily conserved *cis*-acting element in addition to the Krox-20 binding site. *EMBO J* 1996;15:5383–5396.

Vertebrate Limb Development

Cohn, M. J., Patel, K., Krumlauf, R., et al. *Hox9* genes and vertebrate limb specification. *Nature* 1997; 387:97–101.

Hérault, Y., Beckers, J., Kondo, T., et al. Genetic analysis of a *Hoxd-12* regulatory element reveals global versus local modes of controls in the *HoxD* complex. *Development* 1998; 125:1669–1677.

Johnson, R. L., Tabin, C. J. Molecular models for vertebrate limb development. *Cell* 1997; 90:979–990.

Laufer, E., Dahn, R., Orozco, O. E., et al. Expression of *Radical fringe* in limb-bud ectoderm regulates apical ectodermal ridge formation. *Nature* 1997; 386:366-373.

Nelson, C. E., Morgan, B. A., Burke, A. C., et al. Analysis of Hox gene expression with the chick limb bud. *Development* 1996; 122:1449–1466.

Riddle, C., Johnson, R., Laufer, E., Tabin, C. Sonic hedgehog mediates the polarizing activity of the ZPA. *Cell* 1993; 75:1401–1416.

Rodriguez-Esteban, C., Schwabe, J. W. R., De La Pena, J., et al. *Radical fringe* positions the apical ectodermal ridge at the dorsoventral boundary of vertebrate limb. *Nature* 1997; 386:360–361.

Shubin, N., Tabin, C., Carroll, S. Fossils, genes and the evolution of animal limbs. *Nature* 1997; 388:639–648.

Evolution of the Toolkit

The presence of many similar developmental regulatory genes among long-diverged and vastly different animals such as the mouse and the fruit fly raises an apparent paradox. If bilaterians share a related set of developmental genes, how has morphological diversity evolved? There must be genetic *differences* between animal lineages that direct the development of different animal morphologies. One potential source of genetic differences among animals is the coding content of their genomes, including the number and biochemical functions of toolkit genes. Another possible source is the way that toolkit genes are used during development, including both the timing and pattern of their expression and their interactions with other developmental genes in regulatory circuits and networks.

This chapter focuses on the assembly and expansion of the genetic toolkit for development during animal evolution and the role of gene evolution in the origins of morphological complexity. The number of genes contained within an evolving genome is dynamic, and gene duplication increases the information content and potential complexity of developmental programs. Representatives of large gene families are shared among bilaterians, and these families continue to evolve new members in different animal lineages. The expansion of gene families during animal evolution can be traced by comparing the genes found in living (**extant**) organisms and by mapping gene duplication events relative to animal phylogeny. In particular, organisms that are **basal** (most deeply branching) members of a clade indicate the state of the toolkit before the evolution and radiation of certain groups.

Analysis of gene families reveals two periods of major genomic change during animal evolution that correlate with the emergence of more complex animal forms. One interval occurred at the transition to triploblastic bilaterians early in the animal tree; the other is found at the base of the vertebrate lineage. In contrast, the size of the genetic toolkit for development appears roughly equivalent among other morphologically quite diverse bilaterian phyla. Thus expansion of the toolkit correlates with increased animal complexity, but not with diversity.

Evolutionary changes in the proteins that make up the toolkit are often not the primary source of genetic differences that underlie the morphological diversity of animal forms. Indeed, some homologous

> *"Species and groups of species which are called aberrant, and which may fancifully be called living fossils, will aid us in forming a picture of the ancient forms of life. Embryology will often reveal to us the structure, in some degree obscured, of the prototypes of each great class."*
> —Charles Darwin,
> The Origin of Species
> (1859)

genes from long-diverged animal phyla share similar biochemical functions. More significantly, some toolkit genes have similar developmental functions, and these similarities may be used to make inferences about the anatomy of animal ancestors, even without fossil evidence. The shared developmental functions among disparate animal phyla suggest that these toolkit genes controlled the development of various anatomical features in the last common ancestor of bilaterians.

THE HISTORY OF GENE FAMILIES

Conservation of developmental regulatory genes: phylogenetic inferences about animal ancestors

Animal genomes contain many thousands of genes, many of which are required for basic processes common to cellular life. These "housekeeping" genes are shared by most living organisms and predate the evolution of multicellular animal life. Although housekeeping genes are fundamental to cell structure and function, they are not the most likely candidates for the genes critical to the evolution of complex animal body plans. Instead, this chapter focuses on the subset of genes that appears to control patterning and differentiation—namely, the components of the genetic toolkit for development. Previous chapters have described the developmental functions and interactions of many toolkit genes in the model organisms such as *Drosophila* and the mouse. But how and when did this toolkit of genes evolve? And how can we hope to understand the early evolution of toolkit genes, given that the ancient animals that carried those genes are now extinct?

Much of our understanding of the evolution of the genetic toolkit for development is based on deductive logic, using information from extant organisms (primarily the mouse, the fruit fly *Drosophila*, the nematode worm *Caenorhabditis elegans*, the yeast *Saccharomyces cerevisiae*, and some plants) to extrapolate to the past. Similarity between gene sequences found in two (or more) different organisms is most easily explained by common history. In other words, a gene that is conserved among a group of animals must have been present in the last common ancestor of that group. During the subsequent independent evolution of each animal lineage from the last common ancestor, the gene sequence diverged to create related (but not identical) genes, one in each species. Genes with similar sequences that are found within a single animal genome are also related, as they are products of the duplication and divergence of ancestral genes.

Because ancient changes in animal genomes are heritable, the presence of genes in an ancestor can be inferred from genes shared by living taxa. For example, the conservation of toolkit genes in both fruit flies and mice indicates that the genome of their last common ancestor—that is, the common ancestor of all bilaterian phyla (Fig. 4.1)—contained ancestral representatives of all of these shared genes and gene families.

The identification of toolkit genes in different animals (Box 4.1) helps construct the evolutionary history of conserved gene families during the radiation of animal lineages. The characterization of toolkit genes in some organisms is particularly informative. An animal lineage that branches near the base of a clade represents an **outgroup**, which can help to establish the ancestral condition for that clade. For example, the cnidarians are an outgroup to the bilaterian clade, the cephalochordate amphioxus is a close outgroup to the vertebrates, and the onychophora are an outgroup to the arthropods. The resemblance of these animals to their extinct ancestors has led them to be called "living fossils." Their comple-

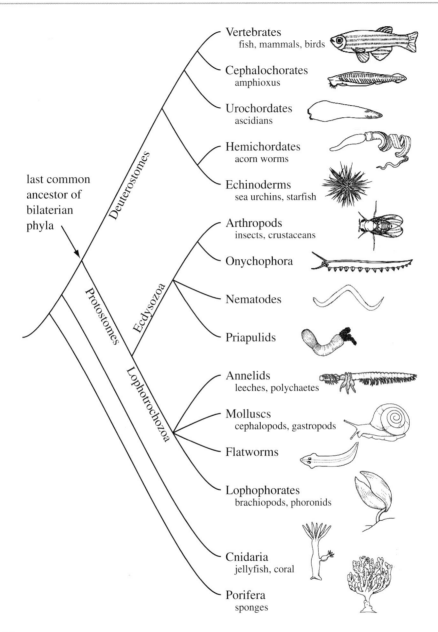

Figure 4.1
Metazoan phylogeny

The phylogenetic relationships between many Metazoan phyla have been resolved using molecular data. Three primary bilaterian clades exist: the deuterostomes (including echinoderms, hemichordates, urochordates, cephalochordates, and vertebrate), the ecdysozoans (including arthropods, onychophora, priapulids, and nematodes), and the lophotrochozoans (including annelids, molluscs, flatworms, and lophophorates). The last common ancestor of all bilaterian phyla is indicated in the figure. Basal branches off of the pre-bilaterian stem lineage lead to the Cnidaria (jellyfish, anemones, coral) and the Porifera (sponges).

Box 4.1 Techniques for Identifying and Analyzing Toolkit Genes in Different Animals

The conservation of developmental regulatory genes allows members of gene families to be identified in different animals based on sequence similarity. Molecular biology techniques such as degenerate polymerase chain reaction (PCR) and library screening facilitate the isolation of homologous genes from both model and nonmodel organisms. Both of these techniques rely on the ability to detect sequence similarity by nucleic acid hybridization between two homologous genes.

Degenerate PCR takes advantage of mixtures of short oligonucleotides that match all possible codon combinations for two conserved peptide sequences within a gene; these mixtures are used to amplify the intervening region from a target pool of nucleic acid (typically, genomic DNA or cDNA). Thus genes that are characterized by conserved protein motifs are good candidates for degenerate PCR, which can rapidly and selectively isolate a region of a gene from any animal's genome.

Library screening, another widely used technique, uses a known gene sequence as a probe to isolate other genes with similar DNA sequence. Libraries are pools of randomly isolated, but unsorted, pieces of genomic DNA or cDNA that can be screened multiple times to identify genes of interest. Such screens are particularly useful for isolating genes from nonmodel organisms.

Today, the identification of toolkit genes in model organisms is, more often than not, an exercise in searching computer databases. The advent of genome sequencing projects (human, fruit fly, nematode, yeast, several plants, and many bacteria) provides large databases for sequence comparison. The number, sequence, and chromosomal map position of members of conserved gene families can be rapidly catalogued based on sequence similarity. Internet-based computer resources, some of which are listed in this box, are the best way to search and analyze genome sequences from model organisms.

The identification of conserved genes from different organisms allows gene sequences to be analyzed for evolutionary relatedness. Many computer programs and tools are available for aligning and comparing the conserved sequences of members of gene families. Large molecular biology Internet servers in the U.S. and Europe include the following:

- The National Center for Biotechnology Information (NCBI) at the National Institute of Heath (*http://www.ncbi.nlm.nih.gov/*)

- The European Bioinformatics Institute, part of the European Molecular Biology Laboratory (*http://www.ebi.ac.uk/index.html*)

- The ExPASy Molecular Biology Server at the Swiss Institute of Bioinformatics (*http://www.expasy.ch/*)

These sites integrate sequence databases (GenBank, SWISS-PROT), protein family databases (PFAM, Prosite), and other genome analysis tools. Powerful computer programs that are used to generate molecular phylogenies ("gene trees") include PHYLIP (*http://evolution.genetics.washington.edu/phylip.html*) and PAUP* (*http://www.sinauer.com/Titles/frswofford.htm*). These programs build molecular phylogenies of related gene sequences using several computer models of molecular evolution.

ment of developmental genes reflects the state of the toolkit before the radiation of bilaterians, vertebrates, and arthropods, respectively. As Darwin foreshadowed (see the opening quote for this chapter), they "aid us in forming a picture of the ancient forms of life."

Gene duplication

One major process involved in the assembly of the bilaterian toolkit for development is the duplication and divergence of genes and the creation of gene families. For example, more than 40% of the genes in the nematode *C. elegans* have sequence similarity to other *C. elegans* genes and thus arose at some point from gene duplication events. Duplicated genes, which are often linked in tandem, may be created by slipping errors during DNA replication, errors in the repair of double-strand DNA breaks, or unequal cross-over events during recombination. Some developmental genes are found as closely related, linked gene pairs, indicating that they derive from tandem duplication events. In *Drosophila*, for example, the gene pairs *engrailed* and *Invected*, *spalt* and *spalt-related*, and *gooseberry* and *gooseberry-neuro* are each tightly linked. Over time, tandemly duplicated genes may become physically separated through chromosomal rearrangements and translocations.

Such tandemly duplicated genes create a unique opportunity for further expansion of a gene family. Unequal crossing over between mispaired tandem copies leads to one chromosome with a duplication and one chromosome with the corresponding deletion (Fig. 4.2). The chromosome that contains the duplicated region also carries a new chimeric gene, located between the parental copies. If more than two tandemly arrayed members

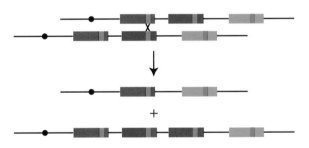

Figure 4.2
Expansion of a tandemly linked gene cluster

Related genes that are tandemly linked present a target for mispairing and unequal crossing over. The sequence similarity between related, linked gene pairs may facilitate mispairing of homologous chromosomes, and a cross-over event might then generate a novel chimeric gene between the mispaired copies. The example shown here is reminiscent of the *Hox* genes, with the conserved homeodomain shown in gray. Crossing over between adjacent paralogous *Hox* genes (shown in red and purple) generates a chimera with the 5′ sequence of one parent and the 3′ sequence of the other, with the break point falling in or near the conserved region. Note that mispairing between more distant genes in the complex can generate duplicated genes in addition to the new chimera. The propensity for the *Hox* genes to remain in a linked cluster may make them particularly susceptible to tandem expansion by this mechanism.

of a gene family exist, unequal crossing over can lead to gene duplication as well as the creation of a new chimeric gene, when the cross-over is out of register by more than one gene. This mechanism of expansion of related tandemly arrayed genes may explain the evolution of the *Hox* complex, for example. New *Hox* genes have evolved as chimeras (or duplicates) of older *Hox* genes, and the *Hox* genes at each end of the complex may represent the "oldest" members of the complex.

Large-scale duplications of chromosomal segments or even of entire genomes (**tetraploidization**) have also occurred during animal evolution. Such duplication events generate large **syntenic** blocks, which are duplicated arrays of genes that are found in the same order, distributed throughout the genome. This mechanism of gene duplication rapidly increases the total number of genes within a genome. The presence of two, three, or four copies of many developmental genes and syntenic regions of linked genes in vertebrate genomes provides evidence of large-scale duplications or tetraploidization events in the vertebrate lineage (Fig. 4.3). The initial tetraploidy of genes and chromosomes may

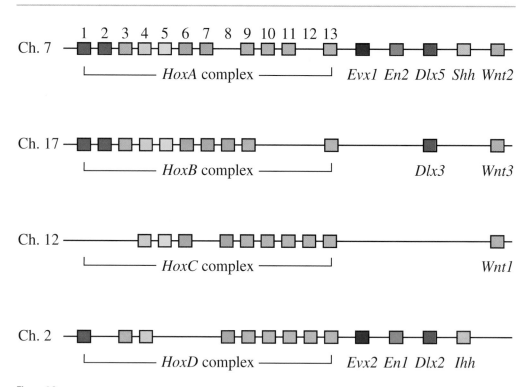

Figure 4.3
Large-scale duplications in vertebrate genomes

Syntenic regions between vertebrate chromosomes are signs of large-scale duplications or tetraploidization events in the vertebrate lineage. Humans have four *Hox* complexes, which are flanked by homologs of the *Evx*, *En*, *Dlx*, *Hh*, and *Wnt* genes (in syntenic order). Some genes have been lost within the *Hox* clusters and, presumably, from the flanking regions of each chromosome (e.g., *Evx* and *En* genes from chromosome 17). Chromosomes are not to scale and intervening genes are not shown.

Source: Redrawn from Postlethwait JH, Yan YL, Gates MA, et al. Nature Genetics 1998;18:345–349.

gradually vanish, as linked genes become separated or are lost due to chromosomal rearrangements and deletions.

Because the duplication and divergence of genes parallels the branching of animal lineages, some terminology has been developed to describe the historical relationships between gene family members found both within and between different animal genomes. All of the genes in a given gene family share sequence similarity and hence are termed **homologs**, as they share common ancestry. Two important distinctions are made among homologous genes, based on how they arise during evolution. Genes that appear in different animals and that arose from a single gene in the common ancestor of those animals are called **orthologs**. For example, the divergence of insects has created orthologous *labial* genes in each insect species. Genes that are related through gene duplication events in a single genome are called **paralogs**. For example, the *Hox* genes of *Drosophila* are a complex of paralogous genes.

The difference between orthologous and paralogous genes is most easily depicted by constructing a phylogenetic tree of related gene sequences. A gene phylogeny depicts the evolutionary relatedness, or history, of members of a gene family. Figure 4.4 shows the relationship between deuterostome *engrailed* genes. Each node on this gene tree represents either an animal lineage bifurcation or a gene duplication. The divergence of animal

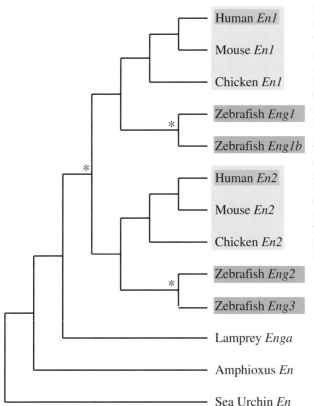

Figure 4.4
Phylogeny of deuterostome *engrailed* genes

This phylogenetic tree depicts the history of the deuterostome *engrailed* genes. All of these *engrailed* genes are homologs, because they share common ancestry. Gene duplication events (indicated by asterisks) have created multiple *engrailed* genes in higher vertebrates. Humans, mice, and chickens have two genes (*En1* and *En2*) and fish have four genes, compared with the single *engrailed* gene found in lamprey, amphioxus, and echinoderms. Yellow boxes represent orthologous genes that arose from animal lineage bifurcations. Blue and green boxes indicate paralogous *engrailed* genes found in the mouse and zebrafish genomes (respectively) that were created by gene duplication.

Source: Redrawn from Force A, Lynch M, Pickett FB, et al. Genetics 1999;151:1531–1545.

lineages created the orthologous *engrailed* genes found in the sea urchin, amphioxus, and the lamprey. A gene duplication generated the paralogous *En1* and *En2* genes found in chicks, mice, and humans. The *En1* genes are more closely related to one another than to any *En2* gene, indicating a more recent common ancestry. Similarly, *En1* genes and *En2* genes are more closely related to each other than to any of the *engrailed* homologs found in more basal vertebrate taxa. Fish *engrailed* genes have been duplicated yet again to create four paralogs—*Eng1a, Eng1b, Eng2,* and *Eng3.*

Mapping gene duplication events onto an animal phylogeny

The relative timing of gene duplication events can be mapped onto an animal phylogenetic tree to provide a complete history of the evolution of a gene family. The presence of the same pair of duplicated genes in multiple animals indicates that the gene duplication event preceded the separation of their respective lineages; thus, this event must have predated their last common ancestor.

Continuing with the analysis of vertebrate *engrailed* genes, it is clear that the *engrailed* gene was duplicated early in vertebrate evolution, as humans, mice, and chicks have two paralogous *engrailed* genes (*En1* and *En2*). The *engrailed* gene phylogeny indicates that one duplication occurred in the stem vertebrate lineage after the divergence of lamprey (Fig. 4.5). A second duplication within the fish lineage generated the four *engrailed* genes found in zebrafish.

One critical step in establishing the relative timing of duplication events in a gene family is the careful choice of one or more outgroups (for example, animal lineages that diverged before a gene duplication) to help determine the ancestral condition for a group of organisms. In the case of *engrailed,* the identification of a single gene in lamprey and in several other basal deuterostome taxa is consistent with the stem deuterostome lineage having a single *engrailed* gene.

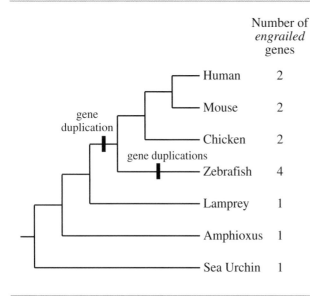

Number of *engrailed* genes

Taxon	Number of engrailed genes
Human	2
Mouse	2
Chicken	2
Zebrafish	4
Lamprey	1
Amphioxus	1
Sea Urchin	1

gene duplication

gene duplications

Figure 4.5

Evolution of *engrailed* genes in deuterostomes

The *engrailed* gene duplication events discussed in Figure 4.4 are mapped onto a deuterostome phylogeny here. The duplication event that created the *En1* and *En2* genes of higher vertebrates appears below the divergence of tetrapods and fish, because both of these genes are present in higher vertebrates, but above the lamprey lineage. Other duplication events in the fish lineage created the gene pairs *Eng1a/Eng1b* and *Eng2/Eng3*, perhaps with a single genome-wide tetraploidization.

In practice, decoding the pattern and timing of gene duplication is often more difficult than suggested by this scenario. The resolution of the branching pattern of gene phylogenies is not always definitive and can be consistent with more than one historical path. It can prove difficult to determine with certainty whether members of a gene family arose from a duplication in a common ancestor or from more recent, independent duplications in different animal lineages. The possibility of gene loss also can obscure the timing of a gene duplication event. It can be difficult to decide whether the absence of a gene reflects gene loss or whether it indicates that the gene never existed in that animal lineage. For example, the first *engrailed* duplication in the vertebrate lineage could have occurred before lamprey diverged, with one copy subsequently being lost in the lamprey lineage. If this scenario is correct, the real timing of the *engrailed* duplication event probably lies deeper within the tree, at the time of large-scale genomic duplications at the base of the vertebrate lineage.

Gene divergence

A gene duplication event may initially generate two redundant gene copies if both the coding sequences and *cis*-regulatory control regions are duplicated. Redundancy can allow one gene copy to be rapidly lost; indeed, gene loss may be a common result following gene duplication. Physical gene loss can occur because of chromosomal deletions, or a duplicated gene can be functionally lost because of the accumulation of deleterious mutations. Large-scale genomic duplications, including tetraploidization, generate large numbers of redundant genes that are not always maintained, as reflected in the loss of some genes from vertebrate *Hox* clusters (see Fig. 4.3). Nonetheless, large gene families of paralogous genes have evolved through the persistence of duplication events. Why, then, are duplicated genes retained within a genome?

Many duplicated genes persist because of functional divergence between the two paralogs. Paralogs can diverge through changes in their coding sequences that lead to differences in protein function, or they may accumulate changes in their *cis*-regulatory elements that generate differences in the timing or pattern of gene expression during development (Fig. 4.6). Duplicated genes do not have to evolve new functions to be retained, if ancestral functions are partitioned between the duplicate copies. Indeed, the modular nature of *cis*-regulatory elements may facilitate the partitioning of ancestral expression domains. For example, one gene copy may lose an element that is retained in the duplicate, whereas the other gene copy loses a different element. In addition, new elements may be individually gained to produce novel expression patterns.

One example of duplicated genes that have diverged primarily in their regulation involves the *Drosophila gooseberry/gooseberry-neuro* gene pair. These linked genes encode functionally redundant *Pax* family transcription factors, yet are expressed in different tissues in early development. The *gooseberry* gene, which is one of the *Drosophila* segment polarity genes, is expressed in stripes in the early embryo. In contrast, *gooseberry-neuro* is expressed at later stages, in the developing nervous system. The expression patterns of these related genes are controlled by different *cis*-regulatory elements. Changes in *cis*-regulation represent the primary evolutionary difference between *gooseberry* and *gooseberry-neuro*.

Assembly of the toolkit: the first animals

The growing list of genes shared by mice and flies (see Chapter 2, and Tables 2.1 and 2.2) reveals that their common ancestor had an extensive toolkit of developmental genes. Basal

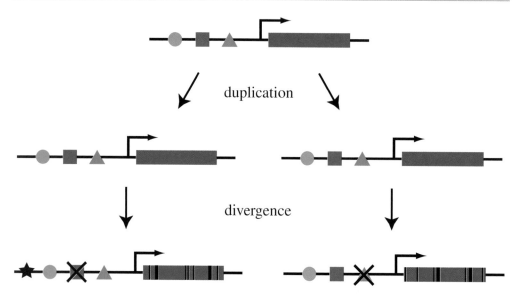

Figure 4.6
Mechanisms of gene divergence

Gene duplication can create two identical copies of a gene, including both *cis*-regulatory regions (blue, red, and green shapes) and the coding sequence (purple rectangle). These duplicated genes can functionally diverge over time in several ways. The coding sequences will accumulate changes (indicated in black), which may alter protein function. The ancestral function may be retained by both proteins, split between them, or retained by a single copy, freeing the other to evolve new functions. In addition, *cis*-regulatory regions may evolve, with separate enhancers acting as independent modules. Over time, an enhancer may be lost in one duplicate, such that the other copy retains that portion of the ancestral expression pattern (indicated by loss of the red square and the green triangle). The duplicated genes can also diverge if new enhancers evolve in one copy (indicated by the black star). In this way, evolution allows the ancestral function of a gene, including both protein function and *cis*-regulatory control, to be shared and even split between duplicated copies, even as new functions and expression domains continue to evolve.

animal lineages, including the diploblast phyla (Cnidaria, Ctenophora) and the Porifera (sponges), have much less developmental and morphological complexity than do bilaterians (see Fig. 4.1). Does their simpler body organization reflect a smaller complement of toolkit genes? Or did the bilaterian toolkit predate the origin and radiation of animals all together?

We can track the *assembly* of the toolkit for animal development by comparing the genes shared among bilaterian phyla to the genes of cnidarians, sponges, and even other eukaryotic organisms. The genome projects for the yeast *S. cerevisiae* and the flowering plant *Arabidopsis*, in addition to the *C. elegans*, *Drosophila*, and human genome projects, have generated data that facilitates the direct comparison of fungal, plant, and animal genomes.

Many protein domains that are characteristic of animal transcription factors or signaling molecules are found in yeast and plants, and thus have ancient origins predating multicellular life (Table 4.1). For example, homeodomains are present in all multicellular organisms, indicating that this protein motif is older than animals themselves. Other domains found in proteins that are crucial to animal development, however, have not been identi-

TABLE 4.1 *Number of genes in shared transcription factor and signaling pathway gene families*

Protein Domain	Fungi S. cerevisiae	Cnidaria and Porifera*	Bilateria		
			C. elegans	*Drosophila*	Human*
DNA Binding					
Zinc finger†	54	?	167	352	≥348
Homeodomain	10	≥19	93	113	≥166
Helix-loop-helix	8	≥1	38	61	≥71
Paired domain	0	≥4	11	13	≥12
T-box	0	≥1	14	8	≥12
Signals/Receptors					
Hh (N-term signal)	0	?	0	1	≥3
Hh (C-term)	‡	?	10§	1	≥3
Dpp/TGF-β	0	≥1	7	6	≥30
Activin receptors	0	?	10	4	≥12
Wg/Wnt	0	?	6	6	≥14
Frizzled	0	≥1	3	7	≥8
Notch ligands	0	?	8	2	≥2
Notch	0	?	2	1	≥2

* Incomplete knowledge of entire genome content; numbers are minimal estimates.
† Classic C2H2-type zinc fingers (transcription factors).
‡ The Hh C-terminal sequence is similar to self-splicing inteins.
§*C. elegans* has several genes with similarity only to the C-terminal sequence of Hh (the *Groundhog* and *Warthog* genes).
Sources: *S. cerevisiae, C. elegans,* and *Drosophila* data from Rubin GM, Yandell MD, Wortman JR, et al. Science 2000;287:2204–2215, supplementary material. Human data from PFAM database at *http://www.sanger.ac.uk/*. Cnidaria/Porifera data from GenBank at *http://www.ncbi.nlm.nih.gov/*.

fied outside the animal kingdom. For example, TGF-β and Wnt signaling molecules, in addition to several families of transcription factors, have been found only in animals (see Table 4.1). Thus some toolkit genes evolved from ancient genes, whereas others appeared early in the animal lineage.

The genes that are conserved between basal animals (Cnidaria, Porifera) and bilaterians represent the minimal genetic toolkit that must have been present in their last common ancestor. Although the genomes of jellyfish and sponges are not as well described as those of model bilaterian organisms, several homologs of developmental genes have been identified, including components of at least one signaling pathway and many transcription factor families (see Table 4.1). Cnidarians have several homeodomain-containing genes that are homologous to bilaterian toolkit genes (including *even-skipped, engrailed, Distal-less,* and *Hox* genes) and at least four members of the *Pax* family of transcription factors. Clearly, at least some members of the bilaterian toolkit were present in the earlier animal lineages.

Although cnidarian genomes contain many of the gene families that contribute to the toolkit, cnidarians appear to have a relatively small number of toolkit genes. For example, bilaterians have at least one additional *Pax* gene and many more *Hox* genes than are found in cnidarians. A comparison of the toolkit genes shared among bilaterians with the genes present in cnidarians indicates that the bilaterian toolkit expanded through gene duplication after the divergence of the cnidarian lineage but before the radiation of bilaterian phyla.

CASE STUDY: EVOLUTION OF THE *HOX* COMPLEX

Expansion of the Hox *complex*

The best-characterized history of a family of developmental regulatory genes is the evolution of the *Hox* genes. *Hox* genes have been isolated from many metazoan phyla, allowing the timing of the tandem duplication events that created the *Hox* complex to be determined relative to the radiation of animal phyla (Fig. 4.7). Given these data, the history of the expansion of the *Hox* complex can be charted through the course of animal evolution.

Cnidarians have two definitive *Hox* genes, which are most closely related to the "anterior" group (*lab/Hox1, pb/Hox2*) and "posterior" group (*AbdB/Hox9-13*) genes found in bilaterians. Although some cnidarians have additional *Hox* genes due to cnidarian-specific duplication events, the common ancestor of Cnidaria and Bilateria most likely possessed only two *Hox* genes. In contrast, all bilaterians have at least two anterior group genes, multiple "central" group genes (*Hox3* through *Hox8*, or their equivalents), and often multiple posterior group genes. Five *Hox* genes (*lab/Hox1, pb/Hox2, Hox3, Dfd/Hox4,* and *Scr/Hox5*), as well as at least one more central gene and one posterior gene, can be clearly identified in all three major bilaterian lineages (the Lophotrochozoa, the Ecdysozoa, and the deuterostomes). This finding indicates that the last common bilaterian ancestor possessed at least seven *Hox* genes. Thus five or more *Hox* genes arose in the bilaterian stem lineage, after the divergence from diploblasts but before the radiation of bilaterian phyla (see Fig. 4.7).

It is important to recognize that not only did several new *Hox* genes arise before the radiation of bilaterian phyla, but the homeodomain sequences of these *Hox* genes also diverged. The various *Hox* paralogy groups can be identified through their characteristic amino acid residues. Because these characteristic residues are conserved in long-diverged bilaterians, they must have been present and fixed in the ancestral *Hox* genes of the last common bilaterian ancestor. Thus the genes of the ancestral *Hox* complex included distinct sequences that have changed very little during the subsequent 500 million years or more of animal evolution.

The *Hox* complex continued to expand during bilaterian evolution. Its evolution in the protostome clade is characterized by sequence divergence of additional central and posterior group *Hox* genes. The two major protostome clades, the Lophotrochozoa (annelids, molluscs, flatworms, lophophorates) and the Ecdysozoa (arthropods, onychophora, priapulids, nematodes), both have central and posterior *Hox* genes specific to their clades. The lophotrochozoans are characterized by the presence of *Lox5, Lox2, Lox4, Post-1,* and *Post-2;* the ecdysozoans are characterized by *Ubx* and *Abd-B* (see Fig. 4.7). These genes may have arisen from independent gene duplication events or from sequence divergence that occurred early in each clade. The respective sets of lophotrochozoan and ecdysozoan

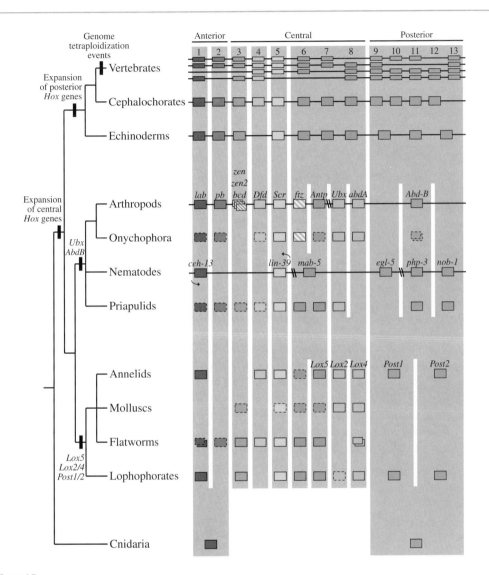

Figure 4.7
Evolution of metazoan *Hox* genes

The relative timing of *Hox* duplication events is mapped onto a phylogenetic tree of Metazoan phyla (**left**), as deduced from the distribution of *Hox* genes in Metazoan phyla (**right**). Cnidarians have *Hox* genes representing only the most anterior and posterior group genes. Early expansion of the *Hox* complex at the base of the bilaterian lineage generated many of the central *Hox* genes. Before the bilaterian radiation, the *Hox* paralogy groups *Hox1-Hox5* were established and fixed. The deuterostome *Hox* complex continued to expand to create *Hox6* through *Hox13*. During early vertebrate evolution, the entire complex became duplicated, thereby creating four complexes in tetrapods (representing the vertebrates). In the ecdysozoan lineage, gene duplications and/or sequence divergence and fixation created the signature ecdysozoan *Hox* genes *Ubx* and *Abd-B*. Similarly, the lophotrochozoans are characterized by *Lox5*, *Lox2*, *Lox4*, *Post1*, and *Post2*. Additional closely related central group genes are present in protostome phyla. Gray-shaded boxes delineate orthology assignments between phyla, when they can be clearly established.

Source: Modified from de Rosa R, Grenier JK, Andreeva T, et al. Nature 1999;399:772–776.

Hox genes were represented in the last common ancestor of each lineage, before the radiation of protostome phyla took place.

The *Hox* complex also expanded early in the deuterostome lineage through tandem expansion of central and posterior group genes. Multiple central and posterior group *Hox* genes have been identified in many deuterostomes, including echinoderms, ascidians (urochordates), and amphioxus (cephalochordate), as well as in vertebrates (see Fig. 4.7). The deuterostome *Hox6-8* (central group) and *Hox9-13* (posterior group) genes do not have clear orthologs in the protostomes, indicating that they arose independently within the deuterostome clade.

Basal deuterostomes have sets of *Hox* genes that are very similar to those found in the vertebrates, but vertebrates have many more genes (see Fig. 4.7). The latter development clearly reflects large-scale duplications or tetraploidization events in the vertebrate lineage that generated at least four separate copies of the entire *Hox* complex. The relative timing of *Hox* complex duplications can be mapped onto chordate phylogeny based on the analysis of *Hox* genes in other deuterostome taxa that represent outgroups to the vertebrate clade. A single *Hox* complex is present in echinoderms, ascidians, and amphioxus (the closest outgroup to the vertebrates). The early vertebrate lineages, including lampreys and sharks, have multiple *Hox* complexes. Therefore, *Hox* complex duplication events map to the base of the vertebrate lineage, after the divergence of cephalochordates (Fig. 4.8). A further round of duplication created additional *Hox* complexes in zebrafish and *Fugu* (see Fig. 4.8). This fish-specific expansion is observed for other genes as well, including the *engrailed* example discussed earlier in this chapter, consistent with an additional large-scale or genome-wide duplication event within the fish clade.

The ParaHox *genes: a sister complex to the* Hox *genes*

The origin of the *Hox* complex is primarily a story of tandem gene duplication and divergence. Nevertheless, additional evidence indicates that the entire pre-bilaterian *Hox* complex of three to four genes was duplicated early in metazoan history. The duplication of

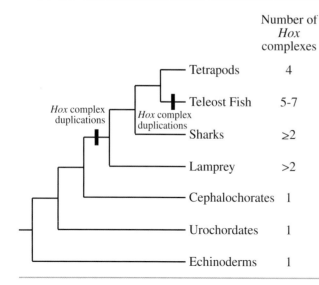

Number of *Hox* complexes

Tetrapods	4
Teleost Fish	5–7
Sharks	≥2
Lamprey	>2
Cephalochorates	1
Urochordates	1
Echinoderms	1

Hox complex duplications

Hox complex duplications

Figure 4.8

Evolution of deuterostome *Hox* genes

The relative timing of vertebrate *Hox* complex duplications are indicated on a phylogenetic tree of deuterostomes. Basal deuterostomes, including echinoderms and cephalochordates, have a single *Hox* complex. At least three complexes have been identified in the primitive jawless lamprey, indicating that *Hox* complex duplication occurred before the divergence of lamprey from higher vertebrates. Teleost fish have undergone an additional round of tetraploidization, creating the seven *Hox* complexes found in zebrafish and at least five in *Fugu*. Tetrapods, including humans, mice, and frogs, have four *Hox* complexes and a total of 39 *Hox* genes.

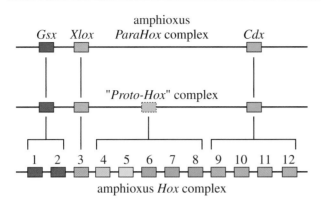

Figure 4.9
ParaHox and ***Hox*** **Complexes**

An ancestral *ProtoHox* complex of three or four genes (**center**) duplicated in early metazoan history to create two sister complexes, the *ParaHox* complex and the *Hox* complex. The number of genes in the *Hox* complex increased through gene duplication and divergence before and during the radiation of bilaterian phyla. Both the *ParaHox* complex (**top**) and the *Hox* complex (**bottom**) are present as tightly linked arrays in *Amphioxus*.

Source: Redrawn from Brooke NM, Garcia-Fernandez J, Holland PWH. Nature 1998;39: 920–922.

this "*Proto-Hox*" complex gave rise to the ancestral *Hox* complex and a sister "*ParaHox*" complex. The *ParaHox* genes (*Gsx, Xlox,* and *Cdx*) have homeodomains with high sequence similarity to the anterior group, *Hox3,* and posterior group *Hox* homeodomains, respectively. The tight chromosomal linkage between these genes in amphioxus (Fig. 4.9) and their colinear deployment in regions of the developing gut are reminiscent of the organization and colinearity of the *Hox* complex. The presence of both *ParaHox* and *Hox* genes in bilaterians and cnidarians suggests that the *Proto-Hox* complex duplication predates the divergence of cnidarians and the stem bilaterian lineage.

New functions for some insect **Hox** genes

The constraint on *Hox* gene sequence and function has been relaxed for some members of the insect *Hox* complex. During insect evolution, two ancestral insect *Hox* genes gained new developmental functions. In the *Drosophila* lineage, duplication and divergence of the *Hox3* gene created the genes *zerknült* (*zen*), *zen2*, and *bicoid* (*bcd*). Also, the *Drosophila fushi tarazu* (*ftz*) pair-rule gene evolved from a central group *Hox* gene that is present in other arthropods.

None of these *Drosophila* genes retains a *Hox*-like role in regulating regional identity along the A/P axis. The *zen* genes are expressed in extraembryonic tissue during early development in several different insects, indicating that the shift in *zen* function occurred fairly early in insect evolution. The *bicoid* gene is critical for the establishment of the A/P axis during *Drosophila* oogenesis (see Chapter 3). The relative timing of the evolution of the *bcd* gene and its role in early establishment of the A/P axis remain unclear, but *bcd* appears to be derived from—and may share ancestral functions with—the *zen* gene. The *Drosophila ftz* gene carries out a pair-rule function during segmentation (see Chapter 3), although this function may represent a more recently evolved role that is not shared among all other insect lineages. The expression of *ftz* in the central nervous system of several insects is more reminiscent of *Hox* genes, with a clear anterior boundary of expression in the thorax. The homeodomains of these genes (*zen, bcd,* and *ftz*) are evolving rapidly, which may reflect a release of functional constraint associated with the evolution of new developmental functions.

The evolution of the *Hox* complex illustrates several important mechanisms underlying the evolution and functional diversification of gene families. First, tandem duplication events expanded the number of paralogous *Hox* genes within the complex. Second, the homeodomains of many *Hox* genes diverged and became constrained before the bilaterian radiation. Third, large-scale or genomic duplications at the base of the vertebrate lineage increased the number of vertebrate *Hox* complexes. Also, an early duplication of the ancestral *Hox* complex created the sister *ParaHox* complex. Fourth, rapid changes in homeodomain sequencing accompanied the evolution of new developmental functions for some insect *Hox* genes.

INTERPRETING THE TOOLKIT: INFERENCES ABOUT ANIMAL EVOLUTION

Expansion of the toolkit and the evolution of morphological complexity

Is expansion of the genetic toolkit for development related to morphological evolution? Two significant periods of expansion of toolkit genes which occurred in the stem lineage of all bilaterians and in the stem lineage of vertebrates, do, in fact, correlate with periods of remarkable increases in developmental and morphological complexity. During the same interval in which the number of *Hox* genes expanded from two to seven distinct genes, the bilaterian stem lineage transitioned from a diploblast body organization to a bilaterally symmetrical triploblast animal. Several dramatic and crucial developmental innovations evolved during this period:

- The establishment of a regulated pattern of early cleavage to form an organized multicellular embryo

- The formation of a continuous gut

- The appearance of a new mesodermal tissue layer with an associated cavity (**coelom**)

- The evolution of distinct anterior–posterior and dorsal–ventral axes

These features are the product of new or modified developmental genetic programs that may well have involved "new" developmental genes.

Large-scale genome duplications at the base of the vertebrate lineage greatly increased the number of *Hox* genes, as well as the number of other toolkit genes. Although these duplications created a great deal of genetic redundancy, many duplicated genes have been retained over a few hundred million years of vertebrate evolution and along the way acquired new or different roles in development. For example, in the mouse, three homologs of the *Drosophila hedgehog* gene (*Sonic hedgehog*, *Indian hedgehog*, and *Desert hedgehog*), are expressed in different tissues and play different roles during development.

During the same evolutionary period, the developmental and anatomical complexity of early vertebrates increased, as reflected by the greater number of differentiated cell types in vertebrates compared to other bilaterians. In particular, the vertebrate central nervous system became larger and more elaborate, which may have contributed to the ecological success exhibited by vertebrates since the Devonian (circa 365 Ma). The developmental complexity and success of the vertebrate lineage may reflect the exploitation of the dramatically larger number of developmental genes in the vertebrate toolkit (see Table 4.1).

Although the expansion of the toolkit may be related to morphological *complexity,* no correlation appears to exist between toolkit expansion and animal *diversity.* The morphological diversity of vertebrates, from humans to hummingbirds, or from whales to snakes, evolved around a common set of developmental genes. For example, mammals, birds, and amphibians share the same set of 39 *Hox* genes. The story is similar for the invertebrate bilaterian phyla, all of which appear to have roughly comparable sets of *Hox* genes. In fact, the same *Hox* genes are found in all arthropods and in the most closely related phylum to the arthropods, the onychophora. The remarkable diversity of fossil and extant onychophorans, trilobites, myriapods, crustaceans, arachnids, and insects evolved around a shared complement of *Hox* genes. The apparent simplicity of some bilaterian body plans, such as the onychophora or the less glamorous flatworms and priapulids, belies the extensive genetic toolkit that these phyla share with more elaborate animal forms.

One interesting exception to the conservation of the toolkit for development is the apparent loss of some toolkit genes in the nematode *C. elegans.* The *C. elegans* genome does not contain orthologs of some genes that are present in both *Drosophila* and vertebrates. For example, *C. elegans* has fewer *Hox* genes than other bilaterians. Because nematodes are ecdysozoans, *C. elegans* must have lost both anterior and central group genes that were present in ecdysozoan ancestors (see Fig. 4.7). Furthermore, *C. elegans* lacks a clear ortholog of a *hedgehog* gene (although this absence may reflect significant divergence of *hh*-like genes in the nematodes). The origin of the *hedgehog* gene predates the bilaterian clade, so the absence (or divergence) of these genes represents a derived condition of *C. elegans.* Many nematodes are parasitic, which may create selective pressures that simplify development and morphology and facilitate the loss (or divergence) of critical developmental genes. Such dramatic gene loss appears to be the exception rather than the rule among bilaterians.

Conserved genes, conserved biochemical functions

The conservation of the genetic toolkit for developmental genes extends beyond primary protein sequences. Even though these genes are used in animals with radically different modes of development, in vivo comparisons of protein activity have revealed similarities in biochemical function between orthologous proteins. In particular, studies have demonstrated the ability of genes from evolutionarily distant species to recapitulate the activity of orthologous genes during development.

For example, overexpression of vertebrate *Hoxb1* and *Hoxb4* genes in *Drosophila* generates phenotypes that resemble the effects of overexpression of the *Drosophila labial* and *Dfd* genes, respectively. The sequence similarity between vertebrate Hox proteins and their arthropod orthologs is limited to the homeodomain and two other short peptides, and these sequences appear to be sufficient for most Hox protein functions. In the case of *Hox* genes and many other toolkit genes, protein sequence conservation often reflects conservation of biochemical functions.

The functional conservation of transcription factors such as the Hox proteins probably derives from the constraints imposed by their regulation of potentially very large numbers of target genes. It may be very difficult for the sequence of a homeodomain to change, as a modification may affect its DNA-binding specificity and therefore alter the ability of the protein to properly regulate all of its target genes. A modification to a Hox protein that simultaneously disrupts multiple Hox-regulated networks would be catastrophic to development. Similarly, signaling pathways are used in a variety of tissues and at different stages of development; hence, a change in the ability of a ligand to interact with its receptor or a change in the interactions between other pathway components could have wide-ranging, dire consequences. For this reason, toolkit proteins are constrained in order to maintain long-established ancestral biochemical functions.

Similarities in the *biochemical* functions of orthologs do not necessarily indicate that these genes are used for the same *developmental* function in their respective organisms, however. This point is most readily apparent for signaling proteins. Conserved biochemical functionality for a signaling protein simply requires conservation of its ability to bind to a receptor and trigger an intracellular signaling cascade. For example, in vertebrates and fruit flies, the Hedgehog protein binds to the Patched receptor protein, thereby activating the Cubitus interruptus/Gli transcription factor (see Chapter 2 and Table 2.2). Nevertheless, the developmental outcome is different in each organism. In the mouse limb bud, the Hedgehog signaling pathway generates different digit identities; in *Drosophila,* this pathway is used during segmentation and later in imaginal discs to establish the A/P compartment boundary. Similarly, orthologous transcription factors may share the same DNA binding specificity and interact with the same co-factors, but perform different regulatory roles during development.

Conserved developmental functions: rebuilding the bilaterian ancestor from phylogenetic inference

Although the conservation of biochemical functionality may simply reflect evolutionary constraint on protein sequence evolution, similarities in the developmental function of homologous genes raise the possibility of deeper evolutionary significance. For example, how should we interpret the striking similarity in the developmental function of *Hox* genes in regulating regional identities along the A/P axis of both mice and *Drosophila?* One possibility is that this similarity is mere coincidence—that the transcription factors independently evolved roles in patterning the A/P axis of protostomes and deuterostomes. More interestingly, the *Hox* genes may have played a role in patterning the A/P axis of the last common ancestor of protostomes and deuterostomes—the hypothetical ancestor of all bilaterians, dubbed "Urbilateria."

One way to distinguish between the possibility that observed similarities are convergent and the likelihood that they reflect conservation of features in a common ancestor is to examine other phyla. The more taxa that share a characteristic, the less likely that the similarity is due to convergence. The deployment of *Hox* genes along the A/P axis of annelids, flatworms, and other bilaterians, for example, suggests that *Hox* colinearity is an ancestral feature of bilaterians.

Not only may we infer the ancestral function of toolkit genes from shared developmental functions among living bilaterians, but we may also infer some of the morphological characteristics of Urbilateria. No fossils have been identified that provide direct knowledge about the morphological complexity of early bilaterian ancestors. Instead, we must use comparative developmental genetic data from living animals in our attempt to rebuild a picture of Urbilateria.

The logic that underlies inferences about the development and morphology of Urbilateria is further illustrated by the conserved role of *Pax6* genes in eye development. Both the mouse *Pax6* gene and the *Drosophila* ortholog *eyeless* are at the top of the regulatory hierarchies that direct eye development in each organism. Other components of the *Pax6-*

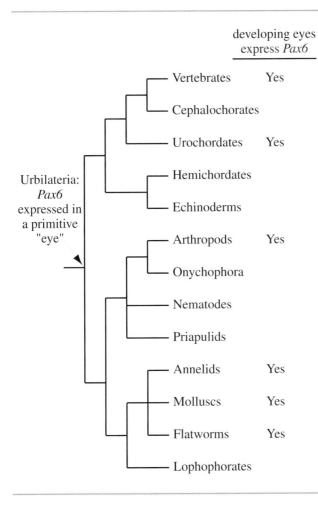

developing eyes
express *Pax6*

Vertebrates	Yes
Cephalochorates	
Urochordates	Yes
Hemichordates	
Echinoderms	
Arthropods	Yes
Onychophora	
Nematodes	
Priapulids	
Annelids	Yes
Molluscs	Yes
Flatworms	Yes
Lophophorates	

Urbilateria:
Pax6
expressed in
a primitive
"eye"

Figure 4.10

***Pax6* expression during eye development in different bilaterian phyla**

Pax6 protein expression have been characterized in many bilaterian phyla. This protein is expressed in the eyes of vertebrates, urochordates, arthropods, annelids, molluscs, flatworms, and nemerteans (not shown). Thus the common bilaterian ancestor may have deployed an ancestral *Pax6* gene during the development of a primitive "eye" or light-sensing organ.

regulated circuit, including the *sine oculis, eyes absent*, and opsin genes, are also shared between flies and mice. Furthermore, *Pax6* orthologs are expressed during eye development in many other bilaterian phyla (Fig. 4.10), providing more evidence that the developmental functional similarities between mouse *Pax6* and *Drosophila eyeless* are not convergent. The remarkable conservation of *Pax6* expression and the *Pax6*-regulated circuit suggests that all bilaterian eyes share a common developmental genetic circuit and that this circuit was present in the bilaterian ancestor. Although we cannot say whether Urbilateria had eyes per se, it may have possessed some type of light-sensing organ whose formation depended on *Pax6* function.

Several other developmental features of Urbilateria may be inferred from the conservation of developmental patterning roles of toolkit genes (Fig. 4.11). The similarity between D/V axis polarization by the *short gastrulation/chordin* genes and TGF-β signaling in insects and vertebrates and A/P axis regionalization by *Hox* genes in many bilaterians suggests that these

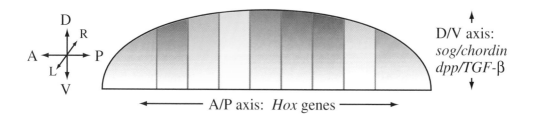

D/V axis:
sog/chordin
dpp/TGF-β

A/P axis: *Hox* genes

+

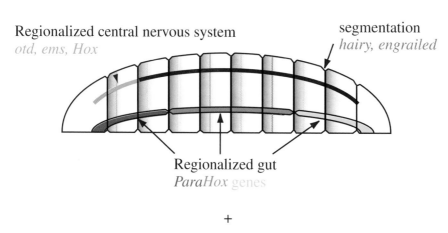

Regionalized central nervous system
otd, ems, Hox

segmentation
hairy, engrailed

Regionalized gut
ParaHox genes

+

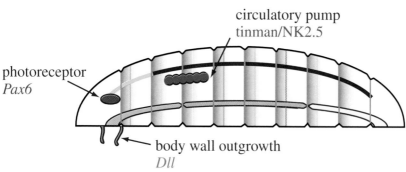

circulatory pump
tinman/NK2.5

photoreceptor
Pax6

body wall outgrowth
Dll

Figure 4.11
Rebuilding "Urbilateria"

The possible features of the common bilaterian ancestor are deduced from the conservation of genes and their developmental functions between arthropods (*Drosophila*) and vertebrates (mouse). (**top**) Patterning of the D/V axis may have been controlled by ancestral genes of the short gastrulation (*sog*)/chordin and TGF-β families. The A/P axis was probably subdivided by nested, overlapping domains of *Hox* gene expression. (**middle**) Different tissue layers were regionally patterned along the A/P axis, including the gut (*ParaHox* genes) and the nervous system [*orthodenticle* (*otd*), *empty spiracles* (*ems*), *Hox*]. Segmentation may have evolved under the regulation of ancestral *hairy* and *engrailed* genes. (**bottom**) Primitive versions of a photoreceptor organ, a circulatory pump, and an outgrowth/appendage might have been present in the bilaterian ancestor, under the regulatory control of the ancestral *Pax6*, *tinman*, and *Dll* genes.

genes also controlled the patterning of the primary body axes of Urbilateria. Using similar logic, we can infer that anteroposterior patterning of the Urbilaterian endoderm, ectoderm, and nervous system may have been regulated by the *ParaHox* genes (*Gsx, Xlox, Cdx*), certain segmentation genes (*hairy, engrailed*), and brain-patterning genes (*emx, otx, Hox*), respectively. Even the expression patterns of the genes involved in dorsoventral patterning of the central nerve cord of flies and mice are similar (*vnd, ind, msh, netrins*). Perhaps most surprisingly, similarities in the genetic regulatory mechanisms that control the development of cardiac tissue (*tinman/Nkx*2.5) and appendages (*Dll*) in insects and vertebrates suggest that Urbilateria possessed primitive versions of these structures.

Without fossils of early animal ancestors, we can only speculate about the complexity of the primitive bilaterian body plan. Perhaps these "organs" were simple structures composed of a few specific cell types. For example, Urbilateria may have had a simple photoreceptor complex rather than an optically sophisticated eye, a contractile muscle regulating hemacoel fluids rather than a modern heart, and a simple outgrowth of the body wall or tentacle-like feeding structure rather than a modern locomotory appendage (see Fig. 4.11). Although the anatomical details remain uncertain, the development of such structures could constrain gene evolution enough to conserve the function of regulatory genes across Bilateria.

All of these features combine to build a fairly complex image of the "primitive" bilaterian, representing an animal that could move (or perhaps crawl) through sediment and could sense and interact with the environment. Perhaps most importantly, this type of ancestor would have many morphological features and genetic tools that might have facilitated a successful evolutionary response to changes in the natural history of early animal life, including climate change and predation.

The revelation of developmental regulatory similarities between long-diverged bilaterians has complicated the assessment of the evolutionary relationships between all of the structures that deploy a shared regulatory gene. For example, it was once believed that "eyes" evolved independently in arthropods, molluscs, and chordates, based on morphological and phylogenetic considerations. The discovery of the role of *Pax6* (and other genes in the circuit) in the development of all sorts of eyes suggests that eyes did not, in developmental genetic terms, evolve repeatedly "from scratch." Similar regulatory circuits are not likely to have been constructed independently by chance, gene by gene, out of the entire repertoire of hundreds of transcription factors in the toolkit.

In another example, the deployment of Dll during the development of body wall outgrowths in all three bilaterians clades (Fig. 4.12) is not likely to be convergent. This similarity, however, does not mean that all bilaterian limbs are homologous—indeed, they are not. All bilaterian limbs did not evolve directly from an Urbilatarian appendage. Rather, they may be considered to be developmental "paralogs" of one another, products of the novel deployment and modification of an ancient and shared regulatory circuit in many different animal lineages over the course of animal evolution.

THE TOOLKIT AS DEVELOPMENTAL POTENTIAL

The deduced complexity of the genome of bilaterian ancestors is much greater than was once thought. Likewise, the image of Urbilateria as an animal with regionally differentiated body axes and a variety of specialized organs is more elaborate than earlier inferences.

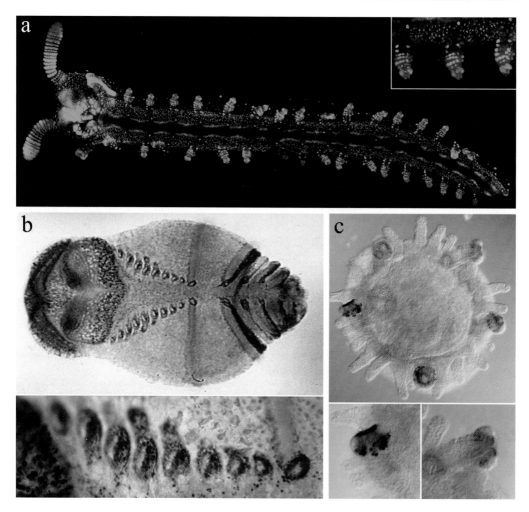

Figure 4.12

Expression of Dll in bilaterian phyla

Dll protein is expressed in body wall outgrowths of many bilaterian phyla, including representatives from all three major bilaterian clades (ecdysozoans, lophotrochozoans, and deuterostomes). (**a**) Dll expression (green) in an onychophoran embryo marks the head appendages and the lobopod walking legs (**inset**). (**b**) In a polychaete (annelid) embryo, Dll is expressed in the growing parapodia (**inset**) and other appendages. (**c**) Dll is expressed in cells at the distal tip of tube feet and spines of a newly metamorphosed sea urchin (echinoderm). The conservation of Dll expression in limbs in diverse phyla reduces the probability that the role of Dll in limb formation reflects co-option or convergence, and suggests that Urbilateria had a primitive limb.

Source: Panganiban G, Irvine SM, Lowe C, et al. Proc Natl Acad Sci USA 1997;94:5162–5166.

These combined genetic and developmental features push the *origin* of animal body patterning deeper into metazoan history. Indeed, the origin of bilaterian complexity—including the evolution of mesoderm, a patterned central nervous system, a regionalized through gut, and primitive eyes, heart, and limbs—corresponds to a period of early animal evolution that was marked by an increase in the number of developmental toolkit genes. The period between the last common ancestor of diploblasts and bilaterians and the radiation of bilaterian phyla represents an interval during which considerable genetic and morphological complexity evolved.

In contrast, the *radiation* of diverse bilaterian body plans is cast against a backdrop of shared genetic tools. Certainly, the genetic toolkit continues to evolve—gene duplication and divergence is a never-ending process. Notably, large-scale duplications early in the vertebrate lineage multiplied the contents of the vertebrate toolkit. Yet the basic components are widely shared among invertebrate bilaterian phyla, and the biochemical functions of the encoded proteins are surprisingly conserved across hundreds of millions of years. What, then, is the genetic basis underlying the morphological diversification of animal forms?

The remaining chapters of this book look not to the evolution of new genes, but rather to higher-order changes in developmental genetic programs. In particular, Chapter 5 focuses on evolutionary changes in gene regulation, including the pattern, timing, and level of gene expression and the genetic wiring of regulatory interactions, which enabled the diversification of arthropod and vertebrate body plans.

SELECTED READINGS

General Reviews

Knoll, A. H., Carroll, S. B. Early animal evolution: emerging views from comparative biology and geology. *Science* 1999; 284:2129–2137.

Scott, M. P. Development: the natural history of genes. *Cell* 2000; 100:27–40.

Slack, J., Holland, P., Graham, C. The zootype and the phylotypic stage. *Nature* 1993; 361:490–492.

Conservation of Developmental Regulatory Genes

Chervitz, S. A., Aravind, L., Sherlock, G., et al. Comparison of the complete protein sets of worm and yeast: orthology and divergence. *Science* 1998; 282:2022–2028.

Rubin, G. M., Yandell, M. D., Wortman, J. R., et al. Comparative genomics of the eukaryotes. *Science* 2000; 287:2204–2215.

Ruvkun, G., Hobart, O. The taxonomy of developmental control in *C. elegans. Science* 1998; 282:2033–2041.

Gene Duplication and Divergence: Assembly and Expansion of the Toolkit

Force, A., Lynch, M., Pickett, F. B., et al. Preservation of duplicate genes by complementary, degenerative mutations. *Genetics* 1999; 151:1531–1545.

Gauchat, D., Mazet, F., Berney, C., et al. Special feature: evolution of Antp-class genes and differential expression of *Hydra Hox/paraHox* genes in anterior patterning. *Proc Natl Acad Sci USA* 2000; 97:4493–4498.

Holland, P. W. Gene duplication: past, present and future. *Sem Cell Develop Biol* 1999; 10:541–547.

Li, X., Noll, M. Evolution of distinct developmental functions of the *Drosophila* genes by acquisition of different cis-regulatory regions. *Nature* 1994; 367:83–87.

Miller, D. J., Hayward, D. C., Reece-Hoyes, J. S., et al. *Pax* gene diversity in the basal cnidarian *Acropora millepora* (Cnidaria, anthozoa): implications for the evolution of the *Pax* gene family. *Proc Natl Acad Sci USA* 2000; 97:4475–4480.

Ohno, S. Evolution by gene duplication. New York: Springer-Verlag, 1970.

Postlethwait, J. H., Yan, Y. L., Gates, M. A., et al. Vertebrate genome evolution and the zebrafish gene map. *Nature Genetics* 1998; 18:345–349.

Case Study: Evolution of the Hox *Complex*

Averof, M., Akam, M. HOM/Hox genes of *Artemia:* implications for the origin of insect and crustacean body plans. *Curr Biol* 1993; 3:73–78.

Brooke, N. M., Garcia-Fernandez, J., Holland, P. W. H. The *ParaHox* gene cluster is an evolutionary sister of the *Hox* gene cluster. *Nature* 1998; 392:920–922.

Dawes, R., Dawson, I., Falciani, F., et al. *Dax*, a locust *Hox* gene related to *fushi-tarazu* but showing no pair-rule expression. *Development* 1994; 120:1561–1572.

de Rosa, R., Grenier, J. K., Andreeva, T., et al. *Hox* genes in brachiopods and priapulids and protostome evolution. *Nature* 1999; 399:772–776.

Falciani, F., Hausdorf, B., Schröder, R., et al. Class 3 *Hox* genes in insects and the origin of *zen*. *Proc Natl Acad Sci USA* 1996; 93:8479–8484.

Finnerty, J. R., Martindale, M. Q. The evolution of the *Hox* cluster: insights from outgroups. *Curr Opin Genetics Develop* 1998; 8:681–687.

Garcia-Fernandez, J., Holland, P. W. Archetypal organization of the amphioxus *Hox* gene cluster. *Nature* 1994; 370:563–566.

Meyer, A., Malaga-Trillo, E. Vertebrate genomics: more fishy tales about *Hox* genes. *Curr Biol* 1999; 7:R210–213.

Conserved Genes, Conserved Functions: Rebuilding the Bilaterian Ancestor

Bodmer, R., Venkatesh, T. V. Heart development in *Drosophila* and vertebrates: conservation of molecular mechanisms. *Develop Genetics* 1998; 22:181–186.

De Robertis, E. M. Evolutionary biology: the ancestry of segmentation. *Nature* 1997; 387:25–26.

De Robertis, E. M., Sasai, Y. A common plan for dorsoventral patterning in Bilateria. *Nature* 1996; 380:37–40.

Halder, G., Callaerts, P., Gehring, W. J. New perspectives on eye evolution. *Curr Opin Genetics Develop* 1995; 5:602–609.

———. Induction of ectopic eyes by targeted expression of the *eyeless* gene in *Drosophila*. *Science* 1995; 267:1788–1792.

Harvey, R. P. *NK-2* homeobox genes and heart development. *Develop Biol* 1996; 178:203–216.

Holland, L. Z., Kene, M., Williams, N. A., Holland, N. D. Sequence and embryonic expression of the amphioxus *engrailed* gene (*AmphiEn*): the metameric pattern of transcription. *Development* 1997; 124:1723–1732.

Kimmel, C. B. Was Urbilateria segmented? *Trends Genetics* 1996; 12:329–331.

Muller, M., Weizsacker, E. V., Campos-Ortega, J. A. Expression domains of a zebrafish homologue of the *Drosophila* pair-rule gene *hairy* correspond to primordia of alternating somites. *Development* 1996; 122:2071–2078.

Panganiban, G., Irvine, S. M., Lowe, C., et al. The origin and evolution of animal appendages. *Proc Natl Acad Sci USA* 1997; 94:5162–5166.

Shubin, N., Tabin, C., Carroll, S. Fossils, genes and the evolution of animal limbs. *Nature* 1997; 388: 639–648.

CHAPTER 5

Diversification of Body Plans and Body Parts

Chapters 2 and 3 described the general principles of the underlying unity in developmental regulatory mechanisms, and Chapter 4 detailed the widespread conservation of the genetic toolkit for development among bilaterians. These chapters set the stage for the consideration of long-standing questions about body plan evolution. Given the extensive genetic similarities of living animals, how did new and vastly different forms evolve from a common bilaterian ancestor? What are the genetic *differences* that underlie the diversity of animal body patterns? This chapter focuses on ways in which evolutionary changes in the regulation of toolkit genes during development contributed to morphological change. Here we examine the relationship between body plan evolution and regulatory evolution, concentrating primarily on the diversification of repeated structures along the primary body axis and of homologous parts between lineages.

The identification of genetic mechanisms underlying body pattern diversity relies on a comparative approach encompassing both model organisms and their relatives. Our understanding of the genetic basis of the radiation of body plans within a phylum is mostly limited to arthropods (including the insect *Drosophila melanogaster*) and chordates (including several vertebrate model systems). Fortunately, these phyla are also exemplary regarding the degree of body plan diversification they display within the framework of a shared body organization. The incredible diversity of extant arthropods, particularly of crustaceans and insects, in combination with the many arthropod and onychophoran forms present in the Cambrian period (530 Ma), provides a rich experimental and historical foundation for a case study of body plan evolution. Similarly, the evolution and subsequent diversification of the axial morphologies of modern chordates provide dramatic examples of large-scale morphological diversification.

Much of our understanding of the role played by regulatory evolution in shaping animal body patterns comes from the *Hox* genes. The *Hox* genes are the best-characterized developmental regulatory genes within and between Metazoan phyla. The genes themselves predate the radiation of bilaterian body plans (see Chapter 4). In this chapter, we examine how *Hox* genes are used during development in related organisms to understand how evolutionary changes in these and other selector genes contribute to body plan evolution. Comparative analyses of *Hox* gene expression domains reveal that major transitions in body organization in both arthropods and tetrapods correlate with shifting spatial boundaries of *Hox* gene expression. In particular, differences in the regulation of *Hox* genes correlate with the diversity of the number and identity of repeated units, such as arthropod and annelid segments and appendages and vertebrate axial elements.

More closely related animal groups with a more conserved body organization, such as insects, exhibit fewer large-scale differences in *Hox* gene expression. The diversification of homologous structures in the context of a stable body plan is largely characterized by regulatory changes downstream of *Hox* (or other selector) gene function. For example, the morphological diversity of insect hindwings and of vertebrate forelimbs is a consequence of evolutionary changes in the assortment of target genes regulated by the *Hox* (or other selector) genes that pattern these appendages. A second mechanism of diversification acting in organisms that share a particular body plan is the modification of *Hox* expression patterns within developmental fields. In this chapter, we discuss the best-understood case studies that illustrate the relationship between morphological diversity and evolutionary changes in the regulation of the *Hox* genes and of their downstream targets.

DIVERSITY OF ANTERIOR/POSTERIOR BODY ORGANIZATION WITHIN ARTHROPODS AND VERTEBRATES

The arthropods are the most successful animal taxa, with the insects alone accounting for roughly 75% of all known animal species. All arthropods share a body plan made up of repeated units—the segments that bear the paired, jointed appendages for which the phylum is named. Different arthropod classes have particular body plans characterized by the subdivision of the body into distinct regions (for example, head, thorax, abdomen) containing specific numbers of segments and by the distribution of appendages on those segments. Thus the dominant theme in the evolution of the various arthropod body plans has been the diversification of segment and appendage number, organization, and morphology (see Chapter 1, Fig. 1.6).

Arthropods include the following groups:

- The myriapods, which have highly repetitive (**homonomous**) trunk segments

- The crustaceans, which display dramatic diversity in segment number and appendage shape and function

- The insects, which possess a stereotypical body organization consisting of a complex head, a thorax bearing six walking legs and two pairs of wings, and a limbless adult abdomen

- The chelicerates, which have an entirely different, two-part body subdivision into prosoma and opisthosoma

In contrast, the sister phylum to the arthropods, the onychophora, consists of animals that are much less diverse and have only a small number of segment and appendage morphologies. These characteristics likely reflect the primitive condition of the arthropod/onychophoran clade.

The arthropod fossil record is rich with diverse arthropod forms, and distinct arthropod and onychophoran body plans are readily apparent in the Cambrian period. The early divergence of this clade remains cryptic, as the segmental morphology and complexity of the earliest (Precambrian) arthropod ancestors have not been documented.

The evolution of different vertebrate body plans is also characterized by the divergence of repeated units—namely, the vertebrae and paired appendages (see Chapter 1, Figs. 1.6 and 1.7). The mesodermal segments of vertebrates (that is, the somites) originate as identical, serially homologous fields that will eventually give rise to the vertebrae and associated processes of the axial skeleton (see Chapter 3). The number and morphology of the vertebrae that constitute each region of the tetrapod vertebral column (for example, cervical, thoracic, lumbar, sacral) differ among the major tetrapod groups. For instance, mammals typically have 7 cervical (neck) vertebrae, as opposed to 13 or more in birds. Snakes can have hundreds of thoracic (rib-bearing) vertebrae. These differences in the number of specific vertebrae are significant in terms of the evolutionary adaptation of the various groups of tetrapods. The elongation of the thorax of snakes and some lizards, coupled with the loss of limbs, allowed them to exploit unique ecological niches. Likewise, the variations in the number of cervical vertebrae in birds and the loss of sacral (tail) vertebrae in some primates are tied to their unique evolutionary histories.

Vertebrate limbs also exhibit a wide range of patterns and functions. The serially homologous forelimbs and hindlimbs of a single animal can have dramatically different morphologies, such as the wings and legs of birds. The shapes and functions of homologous limbs of different animals range from seal flippers to bat wings to human arms.

The evolution of different body plans within a phylum includes both the morphological diversification of repeated body parts within a single animal and the morphological diversification of homologous structures between animal lineages. In genetic terms, serially homologous body parts evolve in the context of the genome of a single species. In contrast, homologous structures that diverge among lineages, such as the forelimbs of different tetrapods or the hindwings of diverse insects, evolve in the context of independently evolving genomes.

Evolution of the genetic control of segmentation in arthropods

When during development does the diversity of arthropod body plans arise? The general similarity of arthropod segmentation suggests that arthropod development may progress through a common regulatory program for segmentation, then diverge to form different patterns of body organization involving different numbers and kinds of segments and appendages. In fact, segmentation in insects, crustaceans, and chelicerates does pass through a stage when the *engrailed* gene is expressed in a stripe in each segment (Fig. 5.1). The *engrailed* gene is a *Drosophila* segment polarity gene—a member of the last tier in the segmentation gene cascade—and operates within cellular fields to establish and maintain segmental boundaries. Earlier stages of development are not so similar among all arthropods, however. In *Drosophila,* segmentation depends on gradients of transcription factor activity in a syncytial embryo; in other arthropods, segmentation occurs over a longer period in a cellularized embryo.

Despite the morphological differences, some earlier steps in the *Drosophila* segmentation cascade appear to be conserved among arthropods. For example, the pair-rule genes *even-skipped, hairy,* and *runt* are expressed in stripes in alternate segments in several insects and

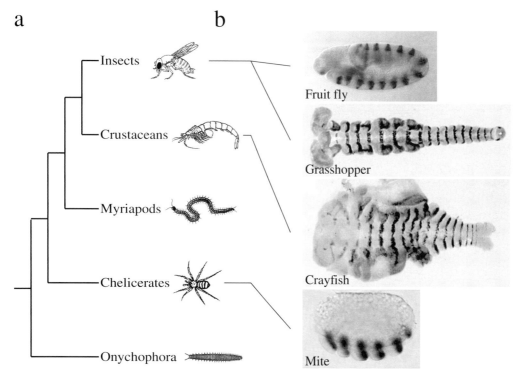

Figure 5.1
Conservation of segmental *engrailed* expression in arthropods

All arthropods share a segmented body plan. (**a**) The phylogeny of major arthropod groups is shown with representative animals.
(**b**) The *engrailed* segment polarity gene is expressed in segmentally iterated stripes during embryogenesis in different arthropods.
This similarity indicates that segmental *engrailed* expression is a conserved stage in the development of all arthropods.
Source: Part b from Patel NH. Development 1994(suppl):201–207; Telford MJ, Thomas RH. 1998;95:10671–10675.

in a spider. Other segmental gene roles, however, are not conserved. For example, the *Drosophila* pair-rule gene *ftz* does not have a pair-rule function in other insects and arthropods. Thus earlier regulatory events in segmentation have diverged between different arthropods, which may reflect the adoption of different modes of early embryogenesis.

Differences evolved in earlier stages of segmentation in two distinct ways. First, genes (such as *Drosophila ftz*) may be independently recruited for a new role in segmentation. Second, the regulatory connections between members of the segmentation cascade appear to be changing.

In the beetle *Tribolium,* for example, the *hunchback* gene has an expression domain similar to the gap expression domain of *Drosophila hunchback,* but the beetle ortholog is regulated differently. The *Drosophila hunchback cis*-regulatory element that directs expression in the gap domain is activated by *bicoid,* whereas the equivalent *Tribolium hunchback cis*-regulatory element is regulated by *caudal*. Thus similar *hunchback* gap expression domains are created by *cis*-regulatory elements that bind different upstream regulators in the beetle and the fruit fly. The genetic regulatory hierarchy controlling segmentation has diverged through the recruitment of additional genes and through regulatory changes within the hierarchy, while the output of *engrailed* expression is conserved among all arthropods. The process of segmentation creates a ground plan upon which the diversity of arthropod segmental identities later develops.

Shifts in trunk Hox *gene expression that mirror changes in arthropod body architecture*

The diversity of arthropod segmental identities suggests that evolution of the *Hox* genes has contributed to the radiation of arthropod forms. These genes have an ancient role in pattern-ing the A/P axis of bilaterians, and they likely regulate the development of distinct segmental identities in all arthropods and onychophora. As transitions in appendage morphology and in body regions occur at different axial positions between the major arthropod groups, the *Hox* genes are key candidates for the genetic players underlying arthropod diversity. Indeed, some of the morphological differences within the arthropod-onychophoran clade correlate with changes in *Hox* gene expression.

We are already familiar with the pattern of *Ubx* expression in *Drosophila,* where the anterior boundary falls within the third thoracic segment and expression extends through most of the abdomen. The anterior boundary of *Ubx* lies near the transition between the thorax, which bears the walking limbs and flight appendages, and the limbless abdomen. Genetic analysis in *Drosophila* has shown that *Ubx* regulates the distinct segmental identi-ties of T3 and A1. In particular, the *Ubx* gene represses leg development in A1 and is responsible for the morphological differences between the wing and the haltere (see Chap-ter 3). An anterior boundary of *Ubx* expression in T3 is characteristic of all insects, reflect-ing the conserved insect body plan.

The anterior boundary of *Ubx* gene expression, however, lies at a different segmental position within the trunks of crustaceans, myriapods, chelicerates, and onychophora (Fig. 5.2). For example, the expression pattern of *Ubx* in a primitive crustacean lineage, the brine shrimp *Artemia,* is quite different from *Drosophila* and other insects (see Fig. 5.2). In *Artemia,* the anterior boundary of *Ubx* falls at the anterior of the thorax, at the transition from the gnathal head segments (which bear feeding appendages) to the thoracic segments (which bear swimming appendages). Thus the expression of this *Hox* gene marks a tran-sition in appendage morphology along the *Artemia* A/P axis, but *Ubx* is expressed at a more anterior position relative to the insects.

Other, more derived crustacean lineages that possess specialized thoracic limbs exhibit different anterior boundaries of *Ubx* expression. In some malacostracan and maxillopod crustacean lineages, one to three pairs of thoracic limbs are reduced in size and are used as feeding appendages (maxillipeds). In these organisms, the anterior boundary of Ubx/abd-A protein expression consistently lies to the posterior of any maxilliped-bearing thoracic seg-ments (Fig. 5.3). Interestingly, in some species, the loss of Ubx/abd-A expression in ante-rior thoracic limbs precedes the developmental transition from a larger limb morphology early in embryogenesis to a smaller maxilliped later in development. In these crustaceans, a shift in *Hox* gene expression anticipates the formation of specialized thoracic limbs.

The myriapods exhibit a body organization that differs from the body organizations of both the insects and crustaceans, and the boundary of Ubx protein expression occurs at a

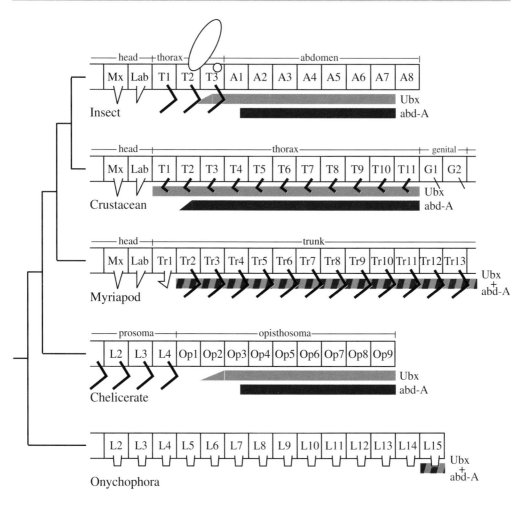

Figure 5.2

Shifts in the anterior boundary of *Ubx* gene expression in arthropods

The expression domains of Ubx (blue) and abd-A (purple) proteins are shown below the body plans of different arthropod groups and onychophora. The anterior boundary of Ubx expression is located in a different segmental position in each group. In each case, this boundary marks a morphological transition in segmental and appendage identity. Diagonal blue and purple stripes in the myriapod and the onychophora indicate the sum of the domains of Ubx and abd-A expression. Segment identities: A (abdominal), G (genital), Mx (maxillary), L (leg), Lab (labial), Lb (lobopod), Op (opisthosomal), T (thoracic), Tr (trunk).

different segmental position. In the centipede, the anterior boundary of Ubx/abd-A expression lies within the second trunk segment and extends through much of the homonomous trunk. Again, this boundary of *Hox* expression marks a transition in limb identity. No Ubx protein is expressed in the developing poison claw on the first trunk segment, whereas it is expressed in all of the developing walking legs on the remaining trunk segments.

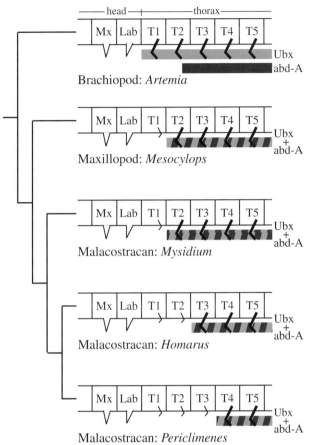

Figure 5.3
***Ubx* expression reflects maxilliped development in crustaceans**

The expression domains of Ubx (blue) and abd-A (purple) proteins are shown below the body plans of different crustaceans. The most primitive crustacean shown (*Artemia;* **top**) expresses Ubx throughout the thoracic segments. Other crustaceans that bear one, two, or three pairs of specialized maxilliped appendages on anterior thoracic segments (shown as smaller thoracic limbs) have progressively more posterior boundaries of Ubx expression. Diagonal blue and purple stripes in Maxilloped and Malacostracan crustaceans indicate the sum of the domains of Ubx and abd-A expression. Segment identities: Mx (maxillary), Lab (labial), T (thoracic).

In the fourth living arthropod class, the chelicerates, the anterior boundaries of *Ubx* and *abd-A* expression fall within the limbless opisthosomal segments.

In summary, the relative positions of the anterior boundaries of *Ubx* and *abd-A* expression have shifted between the various arthropod classes, and even within orders of crustaceans. These shifts correlate with differences in body organization—in particular, with the morphology of thoracic limbs in insects, crustaceans, and myriapods.

In contrast to the highly diversified and patterned segments of arthropods, the onychophoran body plan includes fewer distinct segmental identities and a homonomous trunk. Where are the onychophoran orthologs of the trunk *Hox* genes *Ubx* and *abd-A* expressed? One might have guessed that their deployment would be similar to the arthropods, in which these *Hox* genes are expressed in most of the trunk segments. In reality, Ubx/abd-A protein expression is limited to the extreme posterior of the onychophoran embryo, in the last pair of lobopods and in the terminus (see Fig. 5.2). Although *Hox* genes are conserved between onychophora and arthropods, clearly their expression patterns have changed significantly during their independent evolutionary history.

Hox *genes and the evolution of arthropod heads*

Modern arthropod classes have a more conserved organization of head segments than of trunk segments, with the exception of the chelicerates. Arachnids, for example, have paired walking appendages on the more anterior body region of chelicerates, the prosoma. The prosomal segments exhibit patterns of *Hox* gene expression that correspond to the expression of the anterior *Hox* genes *lab, pb, Dfd,* and *Scr* in crustaceans and insects (Fig. 5.4). In arachnids, these genes have broadly overlapping expression domains, including a segmentally restricted *Hox*-like pattern of expression of the *Hox3/zen* ortholog. These *Hox* gene expression patterns suggests that the chelicerate prosoma is roughly equivalent to the head segments of insects and crustaceans.

The conservation of insect and crustacean head segmentation and patterning correlates with generally similar patterns of expression of anterior *Hox* genes (see Fig. 5.4). The primary exception involves the regulation of the *pb* gene, which is expressed in different segments in insects and crustaceans and even among different insect orders. The insect gnathal head segments where *pb* is expressed form the mouthparts—namely, the mandible, maxilla, and labium. In crickets, beetles, and flies, the maxilla and labium are jointed and bear distal palps (compared with the stubby mandible); *pb* is expressed in both appendages. In contrast, hemipteran insects (including milkweed bugs and bedbugs) are characterized by

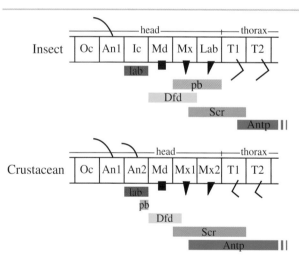
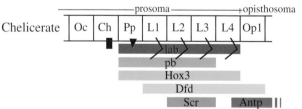

Figure 5.4

Expression patterns of *Hox* genes in the arthropod head

The expression domains of the *Hox* genes *lab, pb, Hox3, Dfd, Scr,* and *Antp* are depicted below the anterior segments of insects (**top**), crustaceans (**middle**), and chelicerates (**bottom**). Chelicerate *Hox* genes have more extensive overlapping expression domains. Some genes, such as *lab* and *Dfd,* have conserved boundaries of expression between these arthropod classes; other genes, such as *pb* and *Antp,* do not. Segment identities: An (antennal), Ch (chelicerae), Ic (intercalary), Md (mandibular), Mx (maxillary), L (leg), Lab (labial), Oc (ocular), Op (opisthosomal), Pp (pedipalp), T (thoracic).

specialized piercing-sucking mouthparts, where the mandibular and maxillary appendages form morphologically similar stylets. The milkweed bug exhibits *pb* expression only in the labium. This more limited pattern of *pb* expression correlates with the coordinated development of the milkweed bug's mandible and maxilla. The evolutionary shifts in *pb* expression observed in insects within the distal gnathal appendages break the usual colinearity of *Hox* expression and likely represent a derived condition relative to other arthropods.

Annelid Hox *expression patterns*

The annelids, or segmented worms, are a second protostome phylum characterized by a body plan made up of repeated units. Whereas the segments of the earthworm and the leech do not have many distinguishing external characters, the polychaete annelids are characterized by diverse segmental identities that correlate with domains of *Hox* gene expression (Fig. 5.5). Annelid *Hox* genes are expressed in a nested, colinear fashion reminiscent of the expression seen in arthropods and vertebrates. Again, the anterior boundaries of orthologous *Hox* genes fall at different segmental positions between annelid lineages. The expression domains of the leech orthologs of *Hox1/lab, Hox4/Dfd,* and *Hox5/Scr* are shifted to the anterior by at least one segment relative to the polychaete *Chaetopterus* (see Fig. 5.5).

Correlation of vertebrate axial patterning with Hox *expression domains*

In vertebrates, changes in the number of vertebrae within regions of the vertebral column correlate with *Hox* expression patterns in the paraxial mesoderm. The transition from one type of vertebra to another corresponds to the anterior limits of expression of specific *Hox* genes (Fig. 5.6). For example, the anterior expression boundary of *Hoxc6* falls at the transition of cervical vertebrae to thoracic vertebrae in mice, chickens, geese, and frogs. Each of these animals possesses a different number of cervical vertebrae, however, and the

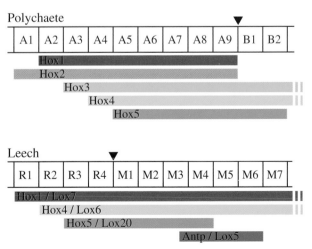

Figure 5.5
Expression patterns of *Hox* genes in annelids

The expression domains of the *Hox* genes *Hox1, Hox2, Hox3, Hox4,* and *Hox5* are depicted below the anterior segments of two annelids. The first nine segments of the polychaete (the A region) bear similar appendages; the B region has a different appendage morphology on each segment. The leech has no appendages, but anterior segments are grouped into R and M regions. Annelid *Hox* genes are expressed in large overlapping domains in polychaetes and leeches, primarily in the nervous system. The anterior boundaries are generally not conserved between leeches and polychaetes, as the leech expression patterns tend to shift to more anterior segments.

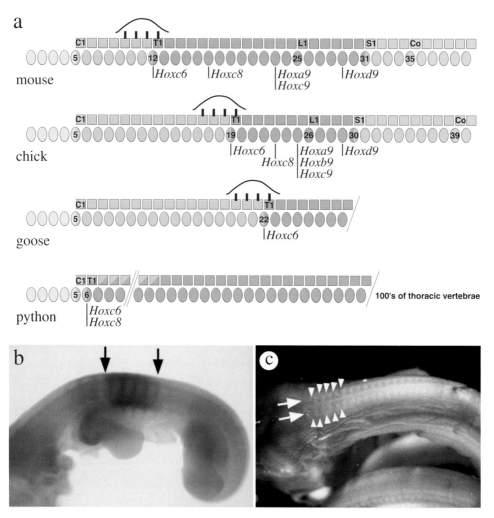

Figure 5.6

Hox **genes and the evolution of tetrapod axial identities**

Differences in the axial organization of tetrapods are reflected in shifts in *Hox* gene expression domains between animals. (**a**) The anterior boundaries of expression of several *Hox* genes in the paraxial mesoderm are shown beneath the somites (circles) and vertebrae (squares) of the mouse, chick, goose, and python body plans. In mammals and birds, which have distinct cervical (green) and thoracic (purple) axial regions, the anterior boundary of the *Hoxc6* gene lies at the cervical–thoracic transition, even though the axial position (somite number) of this transition falls at a different position in each organism. Similarly, the anterior boundary of *Hoxc8* expression lies within the thoracic region of chicks and mice; the *Hoxa9*, *Hoxb9*, and *Hoxc9* boundaries lie near the thoracic–lumbar transition; and the *Hoxd9* boundary lies near the lumbar–sacral transition. In the python, the *Hoxc6* and *Hoxc8* genes have a more anterior expression boundary, reflecting the expanded thoracic vertebral identities of the snake body plan. (**b**) Expression of the mouse *Hoxc8* gene in the thoracic region (the extent of high levels of expression are indicated by arrows). (**c**) Expression of the snake *Hoxc8* gene extends through the anterior of the axial skeleton (indicated by arrows and arrowheads).

Sources: Part a modified from Burke AC, Nelson CE, Morgan BA, Tabin C. Development 1995;121:333–346; parts b and c from Cohn MJ, Tickle C. Nature 1999;399:474–479.

boundary of *Hoxc6* lies at a different axial position relative to the head. In addition, the thoracic-to-lumbar transition, while occurring at different somite levels, appears to be associated with the expression of *Hoxa9, Hoxb9,* and *Hoxc9.* The deployment of these *Hox* genes consistently reflects both the relative position of vertebral identities and the number of vertebrae in each region. These observations suggest that changes in vertebral organization between different vertebrate orders evolved in concert with relative expansions or contractions in *Hox* expression domains.

Another, more dramatic example of shifts in vertebrate *Hox* expression domains is found in snakes. In most tetrapods, *Hoxc6* and *Hoxc8* are expressed in thoracic somites and are required to specify thoracic vertebral identity. The anterior boundary of *Hoxc6* expression marks the cervical-to-thoracic transition; the posterior boundary of *Hoxc8* falls at the position of the hindlimb, just anterior to the lumbar region. The vertebral column of snakes, however, does not have a clear cervical–thoracic boundary. In pythons, expression of *Hoxc6* and *Hoxc8* extends far to the anterior, up to the cranial region (see Fig. 5.6). The posterior boundaries lie at the level of the (vestigial) hindlimb, as in other tetrapods. Within the domain of *Hoxc6* and *Hoxc8* expression, all of the python vertebrae bear ribs, indicating thoracic identity. Interestingly, a subset of cervical characters is present on the most anterior rib-bearing vertebrae of pythons, suggesting that thoracic identities overlie the ancestral cervical identities near the head. Thus the loss of the snake's neck and the expansion of its rib-bearing vertebrae are correlated with the anterior shift in the expression of *Hoxc6* and *Hoxc8.*

How do Hox *domains shift during evolution?*

Each of the bilaterian phyla that have metameric body organization exhibit relative shifts in *Hox* boundaries between lineages. These differences evolved through changes in the regulation of *Hox* gene expression during the radiation of arthropod, vertebrate, and annelid body plans. Mechanistically, shifts in *Hox* expression domains could result from changes in the expression or activity of the proteins that regulate the expression of *Hox* genes. Alternatively, changes could evolve within the *cis*-regulatory regions of the *Hox* genes themselves. The *cis*-regulatory elements that mediate *Hox* expression may evolve to respond differently to upstream regulators (activators or repressors), thereby changing the relative pattern or timing of *Hox* expression.

Circumstantial evidence from comparisons of vertebrate *Hox* genes suggests that evolutionary changes in *Hox cis*-regulation represent diversifying patterns of paralogous *Hox* genes. Vertebrate *Hox* gene homologs that arose from duplications of the ancestral chordate *Hox* cluster sometimes have different boundaries of expression. For example, the chick *Hoxa9, Hoxb9,* and *Hoxc9* genes share a common anterior boundary of expression, but the expression of the *Hoxd9* gene has shifted to the posterior (see Fig. 5.6). These genes evolved from a single common ancestral *Hox9* gene, and they initially had identical *cis*-regulatory regions following cluster duplication events. During their subsequent independent evolution, relative changes in their deployment apparently occurred because of sequence divergence within these *cis*-regulatory regions.

Direct evidence for *cis*-regulatory element evolution has been found for the vertebrate *Hoxc8* gene. The anterior boundary of *Hoxc8* expression lies at a different somite position in mice and chicks, but in a similar location within the thoracic region (see Fig. 5.6). A comparison of homologous *Hoxc8 cis*-regulatory elements in chicks and mice indicates

a

```
mouse         cgtagcc-cagaaatgccacttttatggccctgtttgtctccctgctct-a
baleen whale  ..cg...-.c........t...........g................g..-g
chick         ..c....aa.a.......g......ca..t...........t.....a.gc

mouse         ggttctgaatggggctgaacaaaacagcagtgcagagctggctagacgtct
baleen whale  .a..............c..............c...-..cg............
chick         ..gg......a...gc...........g.ccct...t...............

mouse         gggcttaattgttttatggtttaaataaggtggacactctttcctttga
baleen whale  .....-----.......................................
chick         ..ct.......c.....................gtg.....ct......
```

b

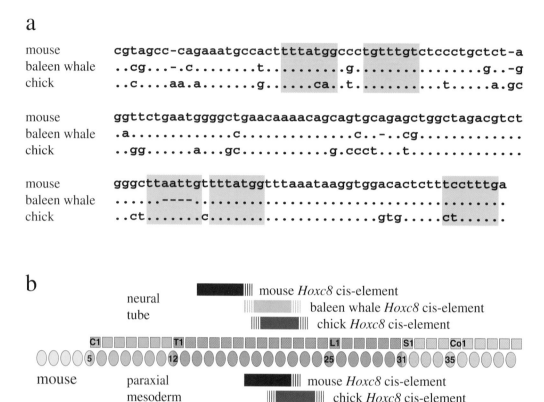

Figure 5.7
Evolutionary changes in the *Hoxc8* early *cis*-regulatory element

The *cis*-regulatory region that controls early axial expression of *Hoxc8* is conserved between mammals and birds. (**a**) Comparison of sequence changes within homologous *Hoxc8* early *cis*-regulatory elements of a mouse, a baleen whale, and a chick. The extensive sequence identity between these homologous elements is indicated by dots, insertion/deletions are indicated by dashes, and sites that are required for function of the mouse *Hoxc8* *cis*-regulatory element are indicated with blue boxes. (**b**) The whale and chick *Hoxc8* *cis*-regulatory elements direct different expression than the mouse sequence, when tested for activity in transgenic mice. The differences in expression boundaries in the neural tube (**top**) and paraxial mesoderm (**bottom**) reflect evolutionary sequence changes in these homologous *cis*-regulatory elements. The shift in the expression domain of the whale element in the neural tube and the loss of paraxial expression are caused by a small deletion, which may be compensated for by other sequence changes in the native whale *Hoxc8* gene. Neural tube expression is shifted to the anterior relative to the paraxial expression, reflecting the equivalent shift in innervation between the spinal cord and the body.

Source: Shashikant CS, Kim CB, Borbely MA, et al. PNAS 1998;95:15446–15451.

that the axial shift in *Hoxc8* expression evolved because of a small number of sequence changes within the highly conserved *Hoxc8* *cis*-regulatory region (Fig. 5.7). The activity of the chick *cis*-regulatory has also been tested in mice, and expression from the chick element appears to be initiated at a more posterior axial position than expression from the mouse element. This shift correlates with the more posterior somite location of the chick thoracic region and the boundary of native chick *Hoxc8* expression. The functional difference between these homologous chick and mouse *Hoxc8* elements suggests that the *Hoxc8 cis*-regulatory region evolved during the divergence of mammals and birds and reflects the different skeletal organizations of these groups.

MORPHOLOGICAL DIVERSITY WITHIN A CONSERVED BODY PLAN

Although some shifts in *Hox* expression correlate well with transitions in axial patterning (as discussed previously for arthropods, vertebrates, and annelids), morphological diversity also occurs among more closely related animals that share a particular body plan. The insects, which generally have conserved boundaries of *Hox* expression domains, exhibit differences in the shape, size, and pattern of their segments and appendages. How have the morphologies of different insects evolved, at a genetic level? The answer to this question, again, lies in evolutionary changes in gene regulation.

Two mechanisms are implicated in the diversification of limb patterning between lineages. First, within a stable pattern of *Hox* gene deployment, regulatory changes downstream of the *Hox* genes have led to the diversification of the shapes and patterns of homologous limbs. A similar process underlies the evolution of morphological differences between vertebrate forelimbs and hindlimbs. Second, some interesting developmental modulations of *Hox* expression patterns within fields have morphological and evolutionary consequences. These upstream differences in *Hox* regulation may provide a glimpse of how larger shifts in *Hox* expression evolve.

Evolution of the limbless insect abdomen

In addition to patterning arthropod appendage morphology, the *Hox* genes played a role in the evolution of the number of limbs in insects. The insect thorax bears three pairs of walking legs and two pairs of wings, but the abdomens of modern adult insects characteristically lack appendages. As many other arthropods have abdominal limbs, the insects are thought to have evolved from an ancestor with limbs on all trunk segments. Some primitive insect fossils have small abdominal leglets, which may represent an intermediary step in the evolution of the limbless insect abdomen. Thus limb development was repressed specifically in the abdominal segments of the insects.

The genetic mechanism underlying the repression of abdominal limbs has been elucidated in *Drosophila*. Abdominal limb development is prevented during embryogenesis via repression of the *Distal-less* (*Dll*) gene in abdominal segments. An early-acting *Dll cis*-regulatory element, which drives expression of this gene in limb primordia in the embryonic thoracic segments, is directly repressed by the Ubx and abd-A proteins. At this stage of embryogenesis, *Ubx* expression does not overlap *Dll* in the third thoracic segment, and *Hox* repression of *Dll* remains limited to the abdominal segments. In *Drosophila,* Hox proteins interact directly with the regulatory network controlling limb development.

In contrast to insects, other arthropods bear limbs on segments that express *Ubx* and *abd-A* (Fig. 5.8). In crustaceans and myriapods, expression of the Ubx/abd-A proteins overlaps with early Dll protein expression in trunk appendages; *Ubx* and *abd-A* do not appear to repress limb development or the *Dll* gene. In collembolans (a primitive insect), three abdominal appendages express Dll in the presence of high levels of Ubx/abd-A. Thus the repression of *Dll* must have evolved in the insect lineage. The evolution of Ubx and abd-A binding sites within *Dll* regulatory elements may have sculpted the limbless abdomen of the insect

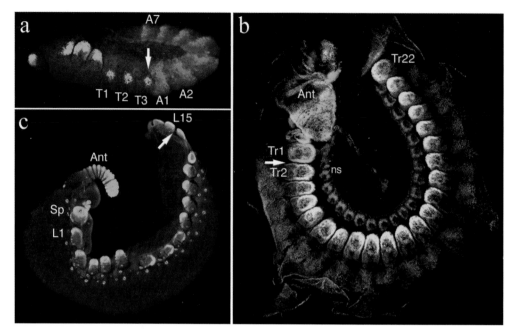

Figure 5.8
Comparison of Ubx and Dll expression in arthropods and onychophora

(a) Ubx (red) and Dll (green) proteins are expressed in separate domains in an early *Drosophila* embryo. Dll is expressed in the limb primordia of head and thoracic (T1–T3) segments, whereas Ubx is expressed in a few cells in T3 and in abdominal (A1–A7) segments. The repression of insect abdominal limbs is associated with the repression of Dll by Ubx and abd-A in *Drosophila*. In other arthropods such as a centipede (**b**) and an onychophoran (**c**), Ubx and Dll protein expression overlaps in some trunk limbs (yellow). The Ubx domain in these animals extends across segments that bear limbs, and there is no evidence that Ubx represses limb development. Segment identities: Ant (antenna), L (lobopod), Sp (slime papillae), T (thoracic), Tr (trunk).

Source: Parts b and c from Grenier JK, Garber T, Warren R, et al. Curr Biol 1997;7:547–553.

body plan.

A comparison of different insect orders indicates that the repression of *Dll* by *Ubx* and *abd-A* may have evolved in two stages. Many insect embryos develop a small appendage on the A1 segment called a pleuropod. In beetles and grasshoppers, Ubx and Dll protein expression overlap during the development of this pleuropod (Fig. 5.9). In fact, *Ubx* is required to properly pattern the pleuropod limb. No limbs develop within the *abd-A* domain, and *abd-A* represses limb development in other abdominal segments. The more primitive condition suggests that *abd-A* repression of *Dll* may have evolved first. Then, in derived insect orders that do not have a pleuropodial appendage on A1 (such as diptera, which includes *Drosophila*), *Dll* became repressed by *Ubx* as well.

Evolution of insect wing number

The scenario depicting the reduction and elimination of limbs on the insect abdomen may be paralleled by the history of the evolution of the number of insect wings. All modern

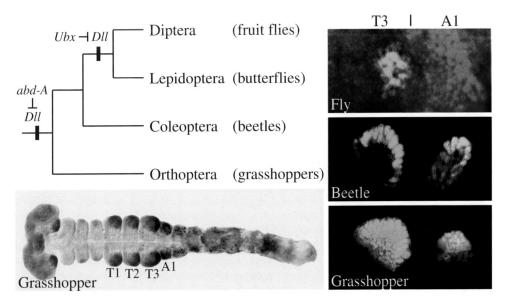

Figure 5.9
Dll and Ubx are co-expressed in the pleuropod appendages of beetles and grasshoppers

The evolution of *Hox* repression of abdominal limb development and *Dll* expression in insects evolved in at least two steps. First, *abd-A* gained the ability to repress limb formation in most abdominal segments (A2–A7) near the base of the insect clade. In the ancestral lineage of butterflies and flies, *Ubx* also evolved the ability to repress limb formation and *Dll* expression in the first abdominal segment (A1). Dipterans do not form an A1 appendage, and Dll (green) is not expressed in that segment (**top right**). In beetles and grasshoppers, a small pleuropod appendage forms in the A1 segment (**bottom right**). Both Dll (green) and Ubx (red) are expressed in these pleuropods (overlap in yellow), indicating that Ubx does not repress pleuropod formation or Dll expression in these animals.
Source: Modified from Palopoli MF, Patel NH. Curr Biol 1998;8:587–590.

winged insects bear two pairs of wings: forewings on the second thoracic segment and hindwings on the third thoracic segment. The invention of wings was a major event in the evolution of insects, catalyzing their radiation and domination of terrestrial habitats.

Insect wings are thought to have evolved only once, early in the insect lineage (the origin of insect wings will be discussed in more detail in Chapter 6). The insect fossil record suggests that, in the primitive state, wing-like structures formed on all of the trunk segments. Fossils of early insects from the Carboniferous (circa 300 Ma) include aquatic larval forms with segmentally iterated, wing-like projections that may have played a functional role in respiration. More recent fossilized mayfly nymphs had larger wing-like structures on thoracic segments and smaller winglets on the abdomen. The fossil record suggests that over time abdominal wing structures became reduced in size and ultimately disappeared, as did the wing-like structures on the first thoracic segment (Fig. 5.10).

The restriction of insect wings to the second and third thoracic segments suggests that the *Hox* genes sculpted the evolution of wing number and pattern. Indeed, genetic analysis has established that specific *Hox* genes repress the initiation of wing development in particular segments. In *Drosophila*, wings are repressed in the first thoracic segment by *Scr* and in the abdominal segments by *Ubx*, *abd-A*, and *Abd-B*. The evolution of *Hox*

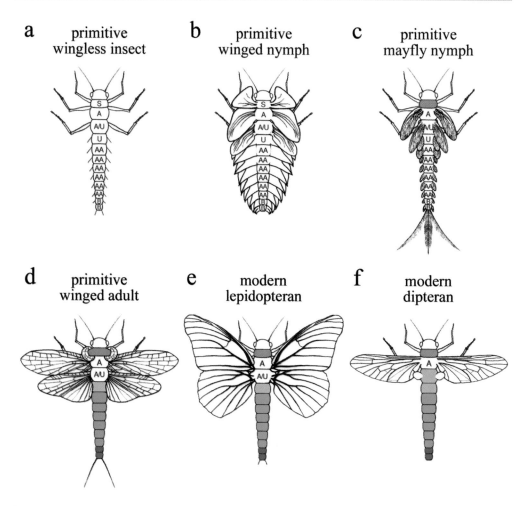

Figure 5.10
Evolution of insect wing number

The segmental domains of *Hox* gene expression did not change during insect evolution and were probably similar in ancestral insects (shown as abbreviations in each body segment: S = *Scr*, A = *Antp*, U = *Ubx*, AA = *abd-A*, B = *Abd-B*). The evolutionary progression of insect wing number reflects the modification or repression of wing development by *Scr*, *Ubx*, *abd-A*, and *Abd-B*. Primitive insects were wingless (**a**), and the first insects with wing-like structures were larval forms (**b, c**). Wing development became repressed in the first thoracic segment (red/*Scr*) and in abdominal segments (yellow/*Ubx*, green/*abd-A*, blue/*Abd-B*) of fossil (**d**) and modern (**e, f**) adult insects. In dipterans (**f**), Ubx also regulates the reduced size and modified shape of the haltere.

Source: Modified from Carroll SB, Weatherbee SD, Langeland JA. Nature 1995;375:58–61.

repression of wing development over the course of insect evolution could explain the reduction and loss of wings on these segments (see Fig. 5.10). Initially, *Scr, Ubx, abd-A,* and *Abd-B* may have modified the expression of wing-patterning genes to control the size of abdominal and T1 winglets. Over time, *Hox* regulation completely suppressed the formation of wings to sculpt the modern insect body plan. The evolution of *Hox* protein binding sites in the *cis*-regulatory regions of genes required for wing formation could account for the repression of wing development during evolution.

Diversification of insect wing morphology

Although all modern winged-insects bear two pairs of wings, many structural, functional, and morphological differences exist between forewings and hindwings both within and between species (Fig. 5.11). For example, the wings of dragonflies (Odonata) appear rather similar, but the forewings of beetles (Coleoptera) have been modified into hardened coverings that protect the hindwings. Butterfly (Lepidoptera) forewings and hindwings are often of similar size, but have evolved different shapes and color patterns. In flies (Diptera), the hindwing has been reduced to the haltere, a small balancing organ with a different biomechanical function.

The differences between forewing and hindwing patterns within species are regulated by the *Ubx* gene. *Ubx* is expressed in the developing hindwing of all insects studied, and it controls hindwing-specific patterning (Fig. 5.12). Yet, the hindwings of different insects such as butterflies, beetles, and flies are all different from one another. This diversity reflects the regulation by *Ubx* of different sets of target genes in each animal.

Recall that in the *Drosophila* haltere, *Ubx* represses the wing-patterning genes that are required for the growth and flattening of the wing and for the development of wing veins and sensory organs (see Chapter 3). In contrast, *Ubx* does not repress the orthologs of these genes in the butterfly. The butterfly hindwing is large and flat, and it has similar veination and sensory organ patterning to the forewing; the developmental program that controls these features is deployed in both wing pairs. In certain butterflies, *Ubx* regulates

Figure 5.11
Diversity of insect wing morphologies

Dragonflies (**top**) have very similar forewings and hindwings, whereas other insects display a diversity of wing morphologies. For example, the forewings of beetles (**middle**) have been modified into specialized protective coverings (elytra). The color patterns of butterfly wings (**bottom**) can differ dramatically between species. (Specimens are not shown to scale.)

Source: Specimens courtesy of Department of Entomology, University of Wisconsin–Madison.

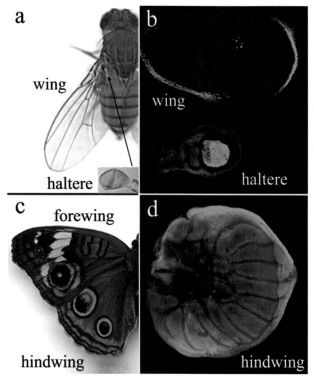

a wing haltere

b wing haltere

c forewing hindwing

d hindwing

Figure 5.12
***Ubx* expression in insect hindwings**

Ubx is expressed in insect hindwings that have diverse morphologies. (**a**) In *Drosophila,* the wing is much larger than the balloon-like haltere. (**b**) Ubx protein (green) is expressed in the developing *Drosophila* haltere, but not in the wing. (**c**) In the butterfly *Precis coenia,* the hindwing is large and flat like the forewing, but has different color patterns and scale morphology. (**d**) Ubx protein (green) is also expressed in the butterfly hindwing.

Source: Modified from Warren R, Nagy L, Selegue J, et al. Nature 1994;372:458–461.

target genes that control the wing shape, color pattern elements, scale pigmentation, and scale morphology—all features that have no counterpart in the fruit fly. Thus the homologous hindwings of flies and butterflies have evolved distinct morphologies in part because of independent changes in the regulatory connections between *Ubx* and genes in the wing hierarchy (Fig. 5.13).

Sets of *Ubx*-regulated target genes may have diverged between species through changes in the *cis*-regulatory elements that control gene expression in the wing field. For instance, the evolution of Ubx binding sites within a wing-specific *cis*-regulatory element could have modified gene expression in the hindwing. Alternatively, the evolution of a new activating Ubx binding site could have created a novel expression pattern in the hindwing. Such evolutionary changes in *cis*-regulatory elements may modify the wing regulatory hierarchy and alter hindwing morphology without globally disrupting wing development.

Modulations in Hox *expression patterns within fields that contribute to insect diversity*

Homologous insect appendages have become diversified in the context of a highly conserved pattern of *Hox* gene expression along the main body axis. Nevertheless, some important differences exist in *Hox* deployment within the homologous limb fields of different insects.

We return to the subject of insect abdominal limbs to discuss an apparent evolutionary **atavism**—that is, reversion to a more ancestral state. Recall that, in general, adult insects

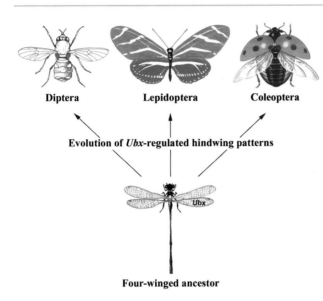

Diptera Lepidoptera Coleoptera

Evolution of *Ubx*-regulated hindwing patterns

Ubx

Four-winged ancestor

Figure 5.13
Evolution of insect hindwing morphological diversity

Evolution of the *Ubx* regulation of hindwing development has contributed to the divergence of insect hindwing morphology. From an ancestral insect with similarly patterned forewings and hindwings (**bottom**), insect hindwings evolved that differ from other homologous hindwings in other species and from the forewings of the same species (**top**). This diversity is associated with evolutionary changes in the set of *Ubx*-regulated target genes in each lineage, in addition to changes in the wing developmental program itself.

Source: Modified from Carroll S. *Nature* 1995;376:479–485.

do not have abdominal limbs. However, butterfly caterpillars (and some other insect larvae) do have abdominal limb-like structures called prolegs, that are used in locomotion. These stubby prolegs are not jointed like thoracic limbs, but their development is marked by *Dll* expression. How does this *Dll* expression escape repression by *Ubx* and *abd-A* in the abdomen?

In early stages of larval development, there is uniform expression of *Ubx* and *abd-A* in the butterfly abdomen and no sign of *Dll* expression or proleg outgrowth. Later, however, the expression of *Ubx* and *abd-A* is turned off within small clusters of cells in the A3–A6 abdominal segments, just before proleg development begins. These cells then begin to express *Dll* and to grow out from the body wall (Fig. 5.14). Thus the apparent evolutionary reversal that allows larval prolegs to develop in lepidopterans (and perhaps in other proleg-bearing species) appears to be a segment-specific modification of abdominal *Hox* gene expression.

A second example of an evolutionary modification of insect limb patterning that is correlated with changes in *Hox* expression within the limb field has been identified in several *Drosophila* species. Subtle differences have been noted in the pattern of tiny leg hairs (trichomes) on the second thoracic legs of *Drosophila melanogaster, Drosophila simulans,* and *Drosophila virilis* (Fig. 5.15). In *D. melanogaster,* trichomes do not form in a certain region of the posterior femur of the T2 leg, resulting in a small "naked valley" of smooth epidermal cuticle. This "naked valley" is associated with high levels of *Ubx* expression, and indeed *Ubx* acts to repress trichome development. *D. simulans* has a larger "naked valley" and high levels of *Ubx* expression over an expanded region of the T2 femur. The more distantly related *Drosophila virilis* lacks a "naked valley" in the T2 trichome array and exhibits only low levels of *Ubx* expression.

Careful genetic dissection of the differences in trichome patterning between *D. melanogaster* and *D. simulans* has indicated that evolutionary differences in the level and extent

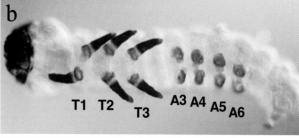

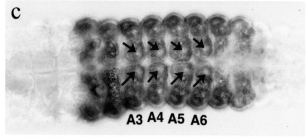

Figure 5.14

Changes in Ubx and Abd-A expression correlate with the development of abdominal limbs in butterfly larvae

Butterfly larvae have abdominal limbs (prolegs) that form on segments A3–A6 (**a**). Dll protein is expressed in all developing larval limbs, including the abdominal prolegs (**b**), suggesting that *Hox* repression of abdominal *Dll* expression has been released during butterfly development. This release from repression is correlated with the disappearance of Ubx and Abd-A proteins from the cells that express Dll and form prolegs in the abdomen (**c**). Abdominal proleg development occurs because of the loss of *Ubx* and *abd-A* expression, rather than because *Hox* repression of *Dll* changes.

Sources: Part b from Panganiban G, Nagy L, Carroll SB. The role of Distal-less in the development and evolution of insect limbs. Curr Biol 1994;4:671–675; part c from Warren R, Nagy L, Selegue J, et al. Nature 1994;372:458–461.

of *Ubx* expression map genetically to the *Ubx* locus. This finding indicates that the source of evolutionary divergence in trichome patterning between *D. melanogaster* and *D. simulans* lies within these fruit flies' respective *Ubx cis*-regulatory control regions. Since the time of the last common ancestor of these species, sequence changes must have evolved in *Ubx cis*-regulatory elements that led to a relative increase in *Ubx* expression in the T2 femur of *D. simulans*. In this way, regulatory evolution at the *Ubx* locus contributed to a quantitative change in T2 leg morphology between *Drosophila* species.

This example demonstrates a potential link between the larger shifts in *Hox* expression seen among distantly related arthropods and the generally similar patterns of *Hox* expression between insects. That is, if even slight differences can evolve in domains of *Hox* gene expression between fairly closely related species, then one can begin to conceive of how larger shifts might arise over greater periods of evolutionary time.

Vertebrate limb diversity: regulatory changes downstream of other selector genes

The vertebrates also possess a wide diversity of limb types, ranging from fish fins to bat wings to human arms. The paired pectoral and pelvic fins of fish are serially homologous

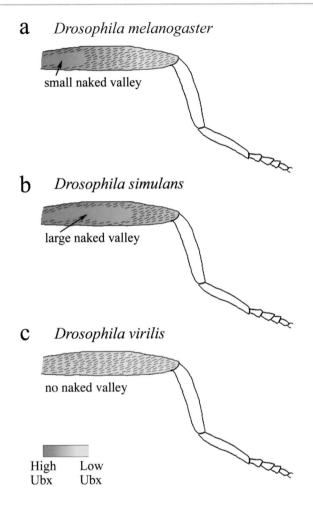

a *Drosophila melanogaster*

small naked valley

b *Drosophila simulans*

large naked valley

c *Drosophila virilis*

no naked valley

High Low
Ubx Ubx

Figure 5.15
Evolutionary changes in Ubx expression regulates morphological differences between related *Drosophila* species

Trichomes (small leg hairs) cover much of the femur of the T2 legs of different *Drosophila* species. A "naked valley" devoid of trichomes appears on the posterior T2 femur of *Drosophila melanogaster* (**a**) and *Drosophila similans* (**b**), but not *Drosophila virilis* (**c**). The size of this "naked valley" in *D. melanogaster* and *D. similans* is controlled by the level of Ubx protein expression in this region during pupation. High levels of Ubx (dark red) repress trichome formation; lower levels do not (orange–yellow). The evolutionary differences in leg morphology and Ubx expression between these species reflect changes in the *cis*-regulation of *Ubx*.

appendages that are homologous to tetrapod forelimbs and hindlimbs, respectively. The evolution of tetrapod forelimb and hindlimb characters is often linked. For example, digits arose at the same time in both sets of paired appendages in Devonian tetrapods. Other, more recent adaptations such as hooves also appeared simultaneously in the forelimbs and hindlimbs of ungulates. Such features likely evolved in concert through the shared developmental genetic pathways that govern the growth and patterning of both limbs (see Chapter 3). The appearance of the same patterning elements in both the forelimb and the hindlimb likely results from the evolution of novel regulatory interactions downstream of a shared limb genetic hierarchy.

Of course, in many familiar cases the serially homologous appendages of vertebrates have evolved independent morphologies. For example, the forelimbs of birds and bats

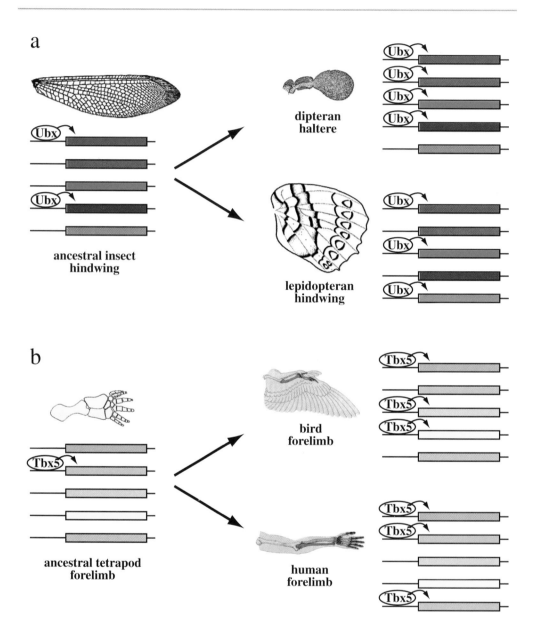

Figure 5.16

Homologous structures diverge through changes in the target genes that are regulated by conserved selector genes

Schematized views of the evolution of insect hindwings (**a**) and vertebrate forelimbs (**b**). The conservation of selector gene expression in insect hindwings (*Ubx*) and vertebrate forelimbs (*Tbx5*) suggests that the ancestral hindwings of insects and ancestral forelimbs of vertebrates also expressed these genes. While the selectors regulated certain target genes (colored boxes) in the ancestral appendage (**left**), these sets of genes changed over the course of evolution, resulting in dramatically different morphologies of homologous limbs in modern lineages (**right**).

were independently modified into flight structures, having very different shapes and functions from the respective hindlimbs of these animals. Kangaroos have small, grasping forelimbs and large, muscular hindlimbs, as do many tetrapods that walk (or jump) primarily on two legs. In chickens and mice, the independent developmental characters of forelimbs and hindlimbs are controlled by limb selector genes that are differentially expressed (see Chapter 3). The same selector genes operate in other vertebrates to regulate the morphological diversification of serially homologous limbs.

The T-box transcription factor *Tbx5* is expressed in the forelimbs of mice, chickens, and newts. A related gene, *Tbx4,* and the homeodomain-containing *Pitx1* gene are expressed specifically in the hindlimbs of these vertebrates. As in the chick and mouse, these genes likely regulate different sets of target genes to pattern morphological differences between the serially homologous forelimbs and hindlimbs of the newt. Furthermore, as the forelimbs (wings) of birds are morphologically distinct from the forelimbs of mammals and amphibians, limb selector genes probably regulate different sets of target genes in these different lineages (Fig. 5.16). Extrapolating to other vertebrates, a common forelimb selector gene, *Tbx5,* probably regulates different target genes in bats and whales and humans.

The story of vertebrate limb diversification both between and within a species is analogous to the regulation of insect hindwing pattern by *Ubx.* In both vertebrates and insects, serially homologous body elements are differentially patterned by field-specific selector genes. Homologous parts between lineages, such as vertebrate forelimbs or insect hindwings, express orthologous selector genes but have diverse morphologies, reflecting evolutionary changes in the array of their respective downstream target genes (see Fig. 5.16).

REGULATORY EVOLUTION AND THE DIVERSIFICATION OF HOMOLOGOUS BODY PARTS

This chapter discussed several case studies of the genetic mechanisms underlying the divergence of arthropod and vertebrate body plans. In each case, morphological change is related to a change in the expression pattern of developmental genes. Shifts in *Hox* expression domains between arthropod, vertebrate, and annelid classes correlate with large differences in body organization—particularly transitions in appendage morphology and vertebral identity. Modifications in *Hox* expression domains within the limb fields of more closely related organisms underlie such morphological features as butterfly prolegs and fruit fly trichome patterning. When *Hox* or other selector gene domains are conserved, evolutionary changes in the regulation of downstream target genes facilitate the diversification of homologous parts between organisms. For example, the *Hox* gene *Ubx* is expressed in all insect hindwings and the limb selector gene *Tbx5* is expressed in all vertebrate forelimbs, yet these genes regulate different sets of target genes in different lineages; the result is insect and vertebrate appendage diversity (see Fig. 5.16).

The flexibility of regulatory networks and the modular nature of *cis*-regulatory control regions allow genetic interactions to evolve without changing the number of genes or even the primary sequence of the proteins encoded by animal genomes. Evolutionary modification of the regulatory interactions between developmental transcription factors and target genes allows different animal morphologies to be built using the same genetic and structural elements. Chapter 6 will extend our discussion of regulatory evolution to include the co-option of developmental regulatory genes in new fields or to pattern new structures. The evolution of novelty will be told as a story of using "old genes to perform new tricks."

SELECTED READINGS

General Reviews

Akam, M. *Hox* genes, homeosis and the evolution of segment identity: no need for hopeless monsters. *Intl J Develop Biol* 1998; 42:445–451.

Carroll, S. Homeotic genes and the evolution of arthropods and chordates. *Nature* 1995; 376:479–485.

Gellon, G., McGinnis, W. Shaping animal body plans in development and evolution by modulation of *Hox* expression patterns. *BioEssays* 1998; 20:116–125.

Arthropod Segmentation

Damen, W. G., Weller, M., Tautz, D. Expression patterns of *hairy, even-skipped,* and *runt* in the spider *Cupiennius salei* imply that these genes were segmentation genes in a basal arthropod. *Proc Natl Acad Sci USA* 2000; 97:4515–4519.

Patel, N. H. It's a bug's life. *Proc Natl Acad Sci USA* 2000; 97:4442–4444.

———. The evolution of arthropod segmentation: insights from comparisons of gene expression patterns. *Development* 1994; (suppl):201–207.

Wolff, C., Schröder R., Schultz, C., et al. Regulation of the *Tribolium* homologs of *caudal* and *hunchback* in *Drosophila:* evidence for maternal gradient systems in a short germ embryo. *Development* 1998; 125:3645–3654.

Arthropod and Annelid Hox *Gene Expression Domains and Segment Identities*

Abzhanov, A., Kaufman, T. C. Homeotic genes and the arthropod head: expression patterns of the *labial, proboscipedia,* and *Deformed* genes in crustaceans and insects. *Proc Natl Acad Sci USA* 1999; 96:10224–10229.

Averof, M., Akam, M. *Hox* genes and the diversification of insect–crustacean body plans. *Nature* 1995; 376:420–423.

Averof, M., Patel, N. H. Crustacean appendage evolution associated with changes in *Hox* gene expression. *Nature* 1997; 388:682–686.

Damen, W. G. M., Hausdorf, M., Seyfarth, E-A., Tautz, D. A conserved mode of head segmentation in arthropods revealed by the expression pattern of *Hox* genes in a spider. *Proc Natl Acad Sci USA* 1998; 95:10665–10670.

Grenier, J., Garber, T., Warren, R., et al. Evolution of the entire arthropod *Hox* gene set predated the origin and radiation of the onychophoran/arthropod clade. *Curr Biol* 1997; 7:547–553.

Irvine, S. Q., Martindale, M. Q. Expression patterns of anterior *Hox* genes in the polychaete *Chaetopterus:* correlation with morphological boundaries. *Develop Biol* 2000; 217:333–351.

Telford, M. J., Thomas, R. H. Expression of homeobox genes shows chelicerate arthropods retain their deutocerebral segment. *Proc Natl Acad Sci USA* 1998; 95:10671–10675.

Vertebrate Hox *Expression Domains and Axial Patterning*

Belting, H-G., Shashikant, C. S., Ruddle, F. H. Modification of expression and *cis*-regulation of *Hoxc8* in the evolution of diverged axial morphology. *Proc Natl Acad Sci USA* 1998; 95:2355–2360.

Burke, A. C., Nelson, C. E., Morgan, B. A., Tabin, C. *Hox* genes and the evolution of vertebrate axial morphology. *Development* 1995; 121:333–346.

Cohn, M. J., Tickle, C. Developmental basis of limblessness and axial patterning in snakes. *Nature* 1999; 399:474–479.

Gaunt, S. J. (2000). Evolutionary shifts of vertebrate structures and *Hox* expression up and down the axial series of segments: a consideration of possible mechanisms. *Intl J Develop Biol* 2000; 44:109–117.

Diversification of Insect Segmental Morphology: Wings and Legs

Carroll, S. B., Weatherbee, S. D., Langeland, J. A. Homeotic genes and the regulation and evolution of insect wing number. *Nature* 1995; 375:58–61.

Palopoli, M. F., Patel, N. H. Evolution of the interaction between *Hox* genes and a downstream target. *Curr Biol* 1998; 8:587–590.

Stern, D. L. A role of *Ultrabithorax* in morphological differences between *Drosophila* species. *Nature* 1998; 396:463–466.

Warren, R., Nagy, L., Selegue, J., et al. Evolution of homeotic gene regulation and function in flies and butterflies. *Nature* 1994; 372:458–461.

Weatherbee, S. D., Nijhout, H. F., Grunert, L. W., et al. *Ultrabithorax* function in butterfly wings and the evolution of insect wing patterns. *Curr Biol* 1999; 9:109–115.

Vertebrate Limb Selector Genes

Logan, M., Tabin, C. J. Role of *Pitx1* upstream of *Tbx4* in specification of hindlimb identity. *Science* 1999; 283:1736–1739.

Niswander, L. Developmental biology. Legs to wings and back again. *Nature* 1999; 398:751–752.

Ruvinsky, I., Oates, A. C., Silver, L. M., Ho, R. K. The evolution of paired appendages in vertebrates *T-box* genes in the zebrafish. *Develop Genes Evol* 2000; 210:82–91.

Tamura, K., Yonei-Tamura, S., Belmonte, J. C. Differential expression of *Tbx4* and *Tbx5* in zebrafish fin buds. *Mech Develop* 1999; 87:181–184.

Weatherbee, S. D., Carroll, S. B. Selector genes and limb identity in arthropods and vertebrates. *Cell* 1999; 97:283–286.

CHAPTER 6

The Evolution of Morphological Novelties

Morphological novelties abound in the history of animal evolution and have been essential to the diversification of animal forms. The scope of novelties encompasses many familiar structures, from antlers to ear bones, from whale tails to panda thumbs. New structures and pattern elements are often key features that distinguish animal groups. For example, the vertebrates are characterized by the forebrain, the neural crest, and cartilage—all novelties that were critical to the success of the vertebrate lineage and that allowed the evolution of further innovations such as teeth and jaws.

Structural novelties (or "key innovations") are associated with adaptive radiations into new ecological territories. The movement of vertebrates onto land is tied to the evolution of the tetrapod limb; the escape of insects into the air required the evolution of wings. The distinctive molar tooth shape has evolved independently several times in association with herbivory. Other novelties such as feathers and butterfly scales permit the display of colors and color patterns used for communication or predator avoidance.

New structures require the evolution of new developmental programs. To understand the origin of morphological novelties, we must look again to the genetic control of development. Have new developmental regulatory genes evolved to sculpt novel body parts? Or are "old" genes recruited for new patterning roles? And, if regulatory genes are reused, are they co-opted individually or as part of larger, preexisting circuits?

This chapter uses case studies to examine various ways in which new structures and pattern elements evolved through changes in developmental gene regulation. In several instances, shared aspects of development and regulatory gene expression reflect the evolution of novelties from preexisting ancestral structures. Such novelties evolved through downstream regulatory changes in developmental programs. In other cases, "old" developmental regulatory genes evolved a new role in the formation of novel pattern elements. In the best-documented examples, groups of genes or entire genetic circuits were recruited to carry out a new developmental function.

> "... on these expanded membranes nature writes, as on a tablet, the story of the modifications of species."
> —Henry Walter Bates, The Naturalist on the River Amazons (1863)

> "Novelties come from previously unseen association of old material. To create is to recombine."
> —François Jacob (1977)

WHAT IS MORPHOLOGICAL NOVELTY?

The history of evolutionary biology is replete with operational definitions of novelty. For our purposes, which are primarily to understand the developmental and genetic basis of the evolution of novel animal forms and patterns, a "novelty" is defined as a structure or pattern element, or even an entire body plan, that has a new adaptive function. This chapter focuses on the best examples of morphological novelty for which developmental genetic knowledge has been elucidated. We do not address other forms of innovation, though they are fascinating in their own right, such as the evolution of physiological adaptations through protein evolution (for example, antifreeze proteins, lens crystallins, keratins, lactose synthesis, immune systems), because they do not concern morphological evolution per se.

Quantitative morphological variation, even when extreme, is not considered novel unless it encompasses a fundamental functional shift. Thus most homologous body parts are not considered novelties, even when the range of a vertebrate forelimb, for example, extends from the gigantic size of a whale flipper to the tiny arm of a tree shrew. Certainly, a gray area exists in terms of the degree of functional change considered sufficient to warrant classification as a novelty. Here, we limit our discussion to cases where a functional shift is accompanied by fundamental changes in development. These examples include the evolution of diverse epithelial appendage types (scales, feathers, and hair), the origin of the insect wing, and the evolution of lepidopteran pigmented wing scales.

Other examples of novelty represent the evolution of a new structure, cell layer, or pattern element with no clear morphological antecedent. The cryptic origins of these innovations make them particularly interesting to evolutionary biologists. Several key innovations evolved in the chordate lineage, including the notochord, the neural crest, and paired limbs. The wealth of developmental genetic knowledge about higher vertebrates and a few key outgroups, such as the cephalochordate amphioxus and urochordate ascidians, allows us to trace the evolution of some chordate novelties through comparisons of these taxa.

This chapter closes by analyzing radical transformations of body organization. Some ascidians have evolved direct development that eliminates the notochord, the defining structure of the chordate lineage. Snakes are "limbless tetrapods," having an extended axial skeleton associated with a novel mode of locomotion. The echinoderms exhibit a novel body plan, characterized by new structural components as well as adult radial symmetry. Although the loss of ancestral characters typically is not considered to be a novelty, in these examples key morphological characters disappear in the context of a new ontogeny.

The origins of most novelties have long puzzled biologists. Observing an anatomically and developmentally complex structure such as a bird feather, a butterfly wing, or the vertebrate brain does not provide insight into the means by which that structure evolved. Nevertheless, taking a comparative approach—integrating the study of paleobiology, comparative embryology, and developmental genetics—provides a window into the history of some important morphological novelties. The evolution of these novelties involves changes in the regulation of transcription factors, signaling molecules, and structural proteins.

NOVEL FUNCTIONS FROM OLDER MORPHOLOGICAL STRUCTURES

Epithelial appendages: scales, feathers, and hair

Scales, feathers, and hair are "epithelial appendages" that cover much of the surface of the skin of reptiles, birds, and mammals. These "appendages" provide protection and are used for thermoregulation and color display purposes. Epithelial appendages have evolved diverse structural and functional properties that disguise their shared evolutionary history. For example, bird feathers have evolved special characteristics adapted for flight, including a lightweight, hollow, branched shaft that allows feathers to interlock. Some mammalian hairs, such as whiskers, are highly sensitive and are used for sensory perception. Both feathers and hairs grow out of basal follicles in the skin that can support the growth of multiple feather or hair morphologies during the animal's life.

The diversity of epithelial appendage morphologies belies the extensive similarities in the initial developmental stages of scales, feathers, and hair. All epithelial appendages originate as epithelial placodes, or small regions of thickened epidermal cells, that will eventually give rise to individual outgrowths (Fig. 6.1a). The spatial distribution of placodes in ordered arrays is established by a series of signals, involving the Hh, Wnt, FGF, Notch, and BMP signaling pathways, between the mesenchymal (dermal) and epithelial (epidermal) tissue layers. The identity of epithelial appendages is determined by the mesenchyme and affected by the timing and level of expression of these signaling molecules. In birds, which have both feathers and scales (on their feet), feathers develop in regions with high levels of Wnt and Notch signaling and low levels of BMP signaling, whereas scales develop in regions with inverse levels of these signals (Fig. 6.1b). Thus modulations in the strength or duration of these early signaling events establish the identities of scales versus feathers. The divergent developmental and structural properties of scales and feathers are determined by subsequent events, such as the deployment of different regulatory genes at later stages of development.

The evolution of diverse epithelial appendage morphologies in different lineages involves changes in the regulatory hierarchies that control epithelial appendage development. One mechanism underlying the diversification of developmental programs within

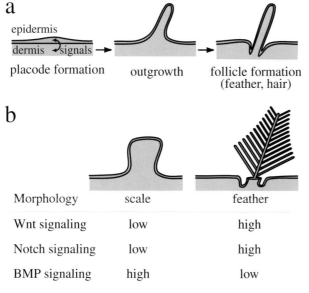

a

epidermis

dermis →signals →

placode formation   outgrowth   follicle formation
(feather, hair)

b

Morphology   scale   feather

Wnt signaling	low	high
Notch signaling	low	high
BMP signaling	high	low

Figure 6.1
Epithelial appendage development

(a) The early stages of epithelial appendage development (scales, feathers, and hair) begin with the formation of an epidermal placode. Signaling between the mesenchyme (dermis) and epithelium (epidermis) establishes the ordered array of placodes on the skin. Outgrowth of the placode is followed by the formation of a follicle, which supports growth and differentiation of feather and hair development. (b) The difference between the identities of avian scales and feathers reflects the timing and level of early signaling events.

lineages involves the recruitment of an existing regulatory gene for a novel patterning function. For example, the *Hoxc13* gene has evolved a new role in epithelial appendage morphogenesis in mice. A novel late pattern of *Hoxc13* expression appears in all developing hair follicles, after the earlier, conserved deployment of *Hoxc13* occurs in restricted axial and limb domains. The mouse *Hoxc13* ortholog is required for proper hair development. Mice that lack a functional *Hoxc13* gene have fragile, brittle hair that breaks off easily. Other epithelial appendages that are hardened by keratin, a structural component of hair, are also affected by *Hoxc13* mutations, including whiskers, nails, and filiform papillae on the tongue. Thus the *Hoxc13* gene appears to have been co-opted to regulate the expression of structural proteins that provide strength and flexibility to mammalian epithelial appendages. The recruitment of regulatory genes for new patterning roles is a common evolutionary mechanism underlying the emergence of several novel structures and pattern elements.

The evolution of insect wings from dorsal leg structures

The evolution of insect wings is recognized as the key innovation behind the amazing evolutionary success of the insect lineage, which accounts for more than 75% of all known animal species. Insect wings arose once during evolution, near the base of the insect lineage. The first wing-like structures, which may have been used for respiration, appeared on each trunk segment of aquatic larval forms in the early Carboniferous. Over the early course of insect evolution, wings became restricted to the adult thorax (see Chapter 5). At first, these structures might have allowed insects to skim across water surfaces; later, they evolved the ability to sustain gliding and powered flight. The use of flight to escape predation, to catch prey, and to move to other habitats catalyzed the evolutionary radiation of the insects.

Interpretations of fossilized insect forms have led researchers to propose two hypotheses concerning the morphological origin of insect wings. One (currently less-favored) hypothesis states that the insect wing evolved from a rigid dorsal outgrowth of the body wall. The second hypothesis suggests that the wing evolved from a dorsal branch (epipodite) of a multibranched crustacean limb (Fig. 6.2a). In the latter scenario, the fusion of the base of the ancestral branched insect limb with the body wall led to the evolutionary displacement of the wing/epipodite away from the rest of the leg. This second model also predicts that developmental regulatory similarities between insect wings and crustacean epipodites should reflect their common evolutionary history.

Indeed, the expression patterns of homologs of two *Drosophila* wing-patterning genes in crustacean appendages are consistent with the origination of insect wings from a dorsal limb branch. In the crayfish, the *pdm* gene is expressed in two distinct domains of the multibranched thoracic limb that are reminiscent of the two patterns of *pdm* (also known as *nubbin*) expression in the fly wing and leg, respectively. The *pdm* gene is expressed throughout a dorsal epipodite in crayfish thoracic limbs, much like the pattern observed in *Drosophila* wing imaginal discs, and in rings in the ventral, walking branch of crayfish thoracic limbs, much like the pattern seen in developing *Drosophila* legs (Fig. 6.2b,c). In the brine shrimp *Artemia*, *pdm* and a homolog of the insect wing dorsal compartment selector gene, *apterous*, are expressed in a dorsal lobe of the thoracic limbs. The parallels between the expression domains of *pdm* and *apterous* in crustacean epipodites and in insect wings suggest that these structures share a common history. This dorsal limb branch is used for respiration and osmoregulation in some extant crustaceans, consistent with the evolutionary origin of insect wings from a gill-like structure with respiratory function.

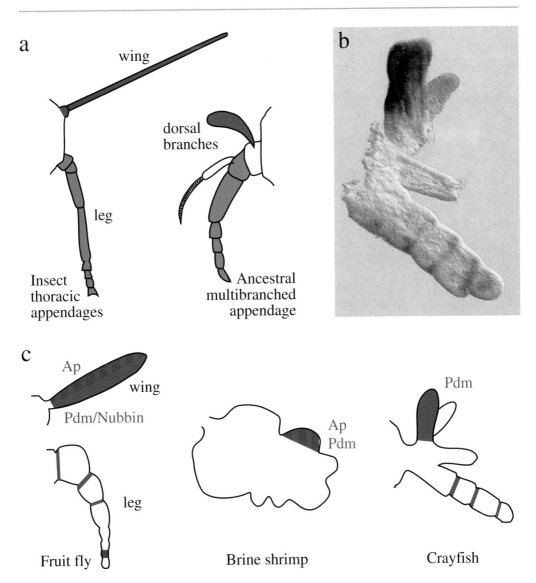

Figure 6.2
Evolutionary origin of the insect wing

The insect wing may have evolved from a dorsal branch of an ancestral multibranched limb. (**a**) The proposed structural homology between the insect wing and an ancestral dorsal leg branch (blue). (**b**) In a crayfish, the Pdm protein is expressed throughout a dorsal branch of the third thoracic limb and in rings in the primary branch (walking leg). (**c**) A comparison of the expression domains of two wing-patterning genes, *apterous* (purple) and *nubbin/pdm* (red), in the thoracic limbs of a fruit fly and two crustaceans. The conserved deployment of these genes in a dorsal branch of crustacean limbs and insect wings supports the evolutionary origin of insect wings from an ancestral branch of the leg.

Source: Part b from Averof M, Cohen SM. Nature 1997;385:627–630; part c redrawn from Jockusch EL, Nagy LM. Curr Biol 1997;7:R358–R361.

Developmental genetic similarities between the crustacean limb branch and the insect wing also suggest that regulatory evolution might explain the morphological differences between these structures. Similarities between the expression domains of the developmental regulatory genes *apterous* and *pdm* imply that downstream changes in target gene regulation have contributed to the divergent morphologies of crustacean limb branches and insect wings. In parallel to the evolution of *apterous*- and *pdm*-regulated patterning, the ancestral multibranched insect limb has become physically divided into two distinct appendages, the wing and the leg, each of which is under independent developmental control.

Butterfly wing color scales: an evolutionary canvas

The evolution of insect wings provided a new patterning surface that has been exploited in many ways by a variety of insect lineages. One of the most striking is the evolution of rows of pigmented scales that cover the wings of butterflies and moths (Lepidoptera). The shingle-like pigmented structural scales form the individual units of butterfly wing color patterns. Scales can exhibit a wide range of colors, shapes, and microarchitectures, both within and between butterfly species. Butterfly scales are also used for thermoregulation and predator avoidance. The array of thousands of overlapping scales on a butterfly wing surface represents a canvas on which a dazzling variety of colors and patterns has evolved.

Butterfly scales represent another morphological novelty that is derived from a preexisting structure—in this case, the innervated bristles that are common to all insects. The early cell lineage of butterfly wing scale development resembles the cell lineage that forms insect bristles (Fig. 6.3a). Each insect bristle consists of four distinct cell types: a bristle cell, a socket cell, a neuron, and a glial (sheath) cell. All four types originate from a single precursor, the Sensory Mother Cell (SMC). The SMC divides to create the pIIa and pIIb cells; pIIa cells then divide again to create a bristle and a socket cell, and pIIb cells divide to create a neuron and a glial cell. In contrast, butterfly structural scales are made up of two cells, a scale cell and a socket cell, and are not innervated. The precursor cell for butterfly scales divides. While one daughter then divides again to create the socket and scale cells (similar to pIIa), the other daughter cell dies.

The evolutionary relationship between butterfly scales and insect bristles is reflected by regulatory gene expression within these cell lineages. The transcription factors that are expressed early in insect bristle development and that are required for bristle formation comprise the bHLH genes of the *Achaete-Scute* Complex (AS-C). A butterfly homolog of these genes (*ASH1*) is expressed both during butterfly sensory bristle formation and in the segregating scale precursor cells that are aligned in rows across the developing wing (Fig. 6.3b). The expression of *ASH1* during butterfly scale formation supports the hypothesis that early scale development is homologous to insect bristle formation.

At least three novel aspects of developmental gene regulation have generated the unique spatial distribution, cytoarchitecture, and pigmentation of butterfly wing scales. One involves the expanded expression of an AS-C gene in rows of cells across the entire surface of the developing butterfly wing. A second regulatory change is the evolution of a set of target genes deployed within scale cells that generates the flattened, rigid cytoarchitecture of butterfly wing scales. A third aspect of regulatory evolution is the recruitment of genes in the pigmentation pathway to be expressed in scale cells, thereby creating an array of colored scales on the adult wing. Collectively, these innovations created a canvas upon which additional regulatory systems control the spatial patterning of colored wing scales.

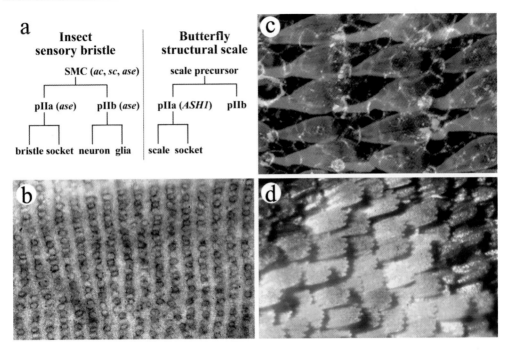

Figure 6.3
Butterfly scales evolved from insect bristles

(**a**) The cell lineage characteristic of insect sensory bristles is similar to the formation of butterfly scales. The butterfly *ASH1* gene, a homolog of the *achaete-scute* bHLH gene family, is expressed in both bristles (not shown) and in developing scale precursor cells (**b**). These cells give rise to the pigmented scales that cover butterfly wings (**c**, pupal, and **d**, adult stages).

Source: Modified from Galant R, Skeath JB, Paddock S, et al. Curr Biol 1998;8:807–813.

Evolution of the butterfly eyespot

The evolution of scale-covered lepidopteran wings provided a new landscape for displaying a striking variety of colors and shapes across the flat wing surface. These color patterns function to warn and confuse predators. For example, a flashing eyespot on a moving butterfly wing can startle a potential predator or redirect its attack away from the butterfly's body. Butterfly eyespots, which are pattern elements composed of concentric rings of colored scales (Fig. 6.4), probably have a single evolutionary origin. Eyespot sizes and colors differ not only between butterfly species, but also between the forewings and hindwings of one animal, and even between the dorsal and ventral surfaces of a single wing. Despite the variation in the color, shape, number, and size of butterfly eyespots, the common features of eyespot development reveal key steps in the evolution of this novel pattern element—specifically, the co-option of genes and regulatory circuits for new patterning roles.

Eyespot development is controlled by a discrete organizer called the "focus," a small group of epithelial cells at the center of the developing eyespot field. During butterfly pupation, focal cells induce the surrounding rings of scale-building cells to produce different pigments. Specification of the eyespot focus is indicated by the expression of the

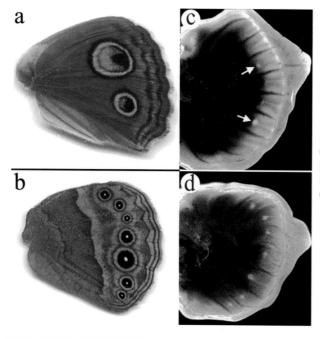

Figure 6.4
Butterfly eyespot foci are marked by Dll expression

The adult hindwings of *Precis coenia* (**a**) and *Bicyclus anynana* (**b**) illustrate some of the variety seen in terms of number and coloration of eyespots between butterfly species. (**c,d**) Dll protein is expressed in larval stages along the wing edge in circular groups of cells of the presumptive eyespot foci. Two foci appear in *Precis* hindwings (**c**, arrows) and seven in *Bicyclus* hindwings (**d**).

Source: Brakefield PM, Gates J, Keys D, et al. Development, plasticity and evolution of butterfly eyespot patterns. Nature 1996;384:236–242.

homeodomain-containing transcription factor *Distal-less* (*Dll*). A novel late pattern of *Dll* expression in focal cells follows an earlier conserved pattern of *Dll* expression (which is found in other insects as well) along the distal edge of the wing. The late pattern begins with the deployment of *Dll* in short rays of cells extending inward from the wing edge in each subdivision of the wing, then resolves to stable expression of *Dll* in circular fields of cells only in the subdivisions in which eyespot foci will form (Fig. 6.4). The continued expression of *Dll* in focal cells in specific wing subdivisions reflects the distribution of eyespots on adult butterfly wings. The conservation of *Dll* expression in the developing eyespot foci of different butterfly species suggests that the novel recruitment of *Dll* expression was an early event in the evolution of eyespots.

The evolution of butterfly eyespots entailed much more than the recruitment of *Dll* expression; indeed, an entire regulatory circuit was co-opted as well. The Hh signaling pathway, which has a conserved function in patterning along the A/P compartment boundary of insect wings (see Chapter 3), has evolved a second wing-patterning role during eyespot formation. Expression of genes in the Hh pathway [*hedgehog* (*hh*), *patched* (*ptc*), and *Cubitus interruptus* (*Ci*)] is modulated in unique patterns within or immediately surrounding eyespot foci (Fig. 6.5). These patterns suggest that this signaling pathway participates in the establishment of the eyespot focus. In addition, a known target gene of the Hh pathway at the wing A/P boundary—the late activation of *engrailed* (*en*) in some anterior compartment cells—is regulated by Hh signaling in the eyespot. En expression is upregulated within every eyespot focus, even in foci that develop in the anterior compartment of the butterfly wing, where *en* is not typically expressed (see Fig. 6.5).

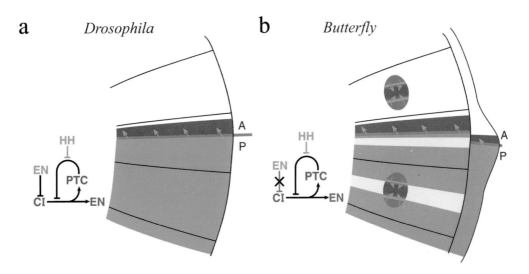

Figure 6.5
Recruitment of the Hh signaling pathway in developing butterfly eyespots

The Hh signaling pathway plays a conserved role in patterning the A/P compartment boundary during insect wing development. In butterfly wings, this pathway is deployed in a novel pattern during the formation of eyespot foci. Co-option of Hh signaling requires changes in the regulation of members of the pathway including the hh and Ci genes (see text for details).

Source: Keys DN, Lewis DL, Selegue JE, et al. Science 1999;283:532–534.

The recruitment of the Hh signaling pathway in the eyespot field is novel in two respects. First, the modulation of *hh* expression in a novel pattern near developing eyespots is a unique feature of *hh* expression in the butterfly wing. Second, the expression of *ptc* and *Ci* in the posterior compartment and of *hh* in the anterior compartment of butterfly wings deviates from the compartmentally restricted deployment of these genes in the *Drosophila* wing. In *Drosophila*, expression of *ptc* and *Ci* is restricted to the anterior compartment because *en* represses the expression of these genes in the posterior compartment. Consequently, *en* repression of *ptc* and *Ci* must be relieved in butterfly wings so that the Hh pathway can be used in the posterior compartment. The combination of modulated *hh* expression and release of the repression of *ptc* and *Ci* allows the Hh pathway to be activated in each developing eyespot field. The novel deployment of *hh*, *ptc*, *Ci*, and *en* in developing eyespots probably represents a critical subset of the regulatory changes that led to the co-option of a complete signal transduction circuit.

It has been proposed that eyespots evolved from simpler spot patterns of uniform color. The expression of Dll and En in the centers of eyespots appears to be a feature shared among eyespots and may offer a clue to eyespot origins. Namely, these genes may have been expressed in simpler, uniform spot patterns. The evolution of multi-ringed eyespots may then have involved the acquisition of signaling activity by cells expressing these genes

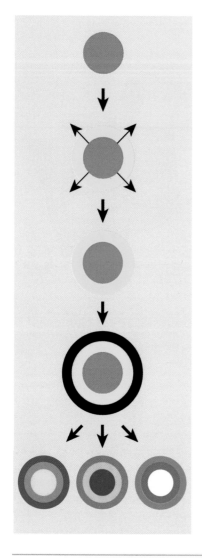

Gene expression
in spot (center)
(e.g. *Dll*, *en*)

Evolution of
signaling activity

Recruitment of
new target genes

Recruitment of
additional genes

Divergence of
pigmentation
gene regulation

Figure 6.6

The evolution of eyespots from simple spots

It is thought that eyespots evolved from simpler spots of uniform color. Dll and En may have been expressed during the development of these primitive spots. As cells within this spot acquired signaling activity, surrounding cells acquired the ability to express different pigmentation genes. Further recruitment of signal-regulated genes led to evolution of multi-ringed eyespots. Different responses to focal signaling among species may underlie different eyespot color schemes.

and the recruitment of genes induced by this new signaling activity (Fig. 6.6). The diversity of eyespot color schemes may have evolved through evolutionary changes in genetic responses downstream of signaling from the focus (Fig. 6.6).

The development and evolution of butterfly eyespots illustrate two recurring themes in the evolution of novel characters. First, conserved regulatory circuits can be recruited for new roles during the development of novel morphologies. This recruitment requires changes in the regulation of at least one gene, and subsequent deployment of other genes in the circuit may result from existing regulatory linkages, such as a signal transduction pathway. In this way, a large suite of genes may be deployed in a novel structure with just a small number of evolutionary regulatory changes. Second, evolutionary changes in target gene regulation can facilitate morphological diversification of a novel character. As regulatory evolution modifies the genetic interactions within a developmental program, new patterns can emerge both within and between species.

THE EVOLUTION OF VERTEBRATE NOVELTIES

The success of the vertebrates can be attributed to several unique traits that arose during the evolution of vertebrates from a deuterostome ancestor. A limited list of novelties that evolved in the vertebrate lineage includes the following:

- Paired pectoral and pelvic limbs
- A large, regionalized, and complex brain
- A migratory population of neural crest cells
- A notochord

By comparing the morphology, ontogeny, developmental genetics, and fossil history of vertebrates and other chordates (such as the cephalochordate amphioxus and urochordate ascidians), it is possible to trace the evolutionary origins of these structures. Conserved regulatory genes control the pattern of these new developmental fields and tissues, revealing once again that new developmental programs have evolved through the co-option of existing regulatory genes and circuits and the expansion of ancestral patterning roles.

Fins to limbs: paired appendages and the tetrapod hand

The adaptive evolution of vertebrates capable of surviving in aquatic, terrestrial, and aerial environments involved the acquisition and modification of paired pectoral and pelvic appendages. These limbs boosted vertebrates' maneuverability and speed in water and later were used as the primary means of locomotion on land. The early history of paired appendages remains uncertain, however. Diverse fin morphologies are present in Ordovician fish (463–439 Ma). Some of these ancient fish possessed median fins running the length of the body, some had paired fins near the head, and others had no fins at all. Based on the fossil record, median fins appear to have preceded the appearance of paired fins, and paired pectoral fins preceded the evolution of pelvic fins (Fig. 6.7).

The group of bony fishes that later gave rise to the tetrapod lineage included animals with similarly patterned pectoral and pelvic fins. This similarity may reflect the deployment of the same regulatory genes in each limb pair. Modern tetrapods and bony fish exhibit similar patterns of *Hox* gene expression during the development of pelvic and pectoral appendages. The posterior genes of both the *HoxA* and *HoxD* complexes (*Hox9-13*) are expressed in complex and dynamic patterns during tetrapod limb development (see Chapter 3). In *unpaired* fish fins, however, these *Hox* genes are not deployed. Consequently, primitive vertebrate appendages may not have been patterned by *Hox* genes, and *Hox* genes may have been recruited later to pattern paired fins.

The evolution of posterior *HoxA* and *HoxD* gene expression in both pectoral and pelvic limbs may have occurred in two discrete steps (see Fig. 6.7). First, axial patterns of posterior *Hox* gene expression may have been recruited during the development of the adjacent paired pelvic limbs. Later, the pectoral limbs may have co-opted the pelvic pattern of *Hox* deployment. The nested domains of *Hox* gene expression found along the primary body

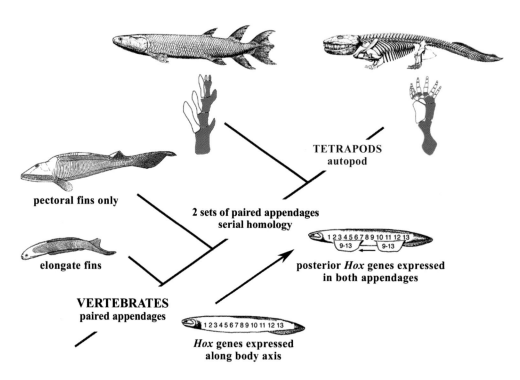

Figure 6.7
Evolution of vertebrate paired appendages

Vertebrate paired appendages evolved in a series of steps. Paired appendages are first encountered in the fossil record as elongate fins extending laterally along the body wall or as paired pectoral fins in jawless fish (shown on basal branches). Jawed fish are characterized by multiple sets of paired appendages that often have a defined fin axis (shown in blue; **top left**). Tetrapods (**top right**) have a distal autopod, with digits branching from a bend in the limb axis. One step in the evolution of serially homologous paired appendages was the co-option of nested patterns of posterior *Hox* genes expression in both sets of paired appendages (**lower right**).

Source: Modified from Shubin N, Tabin C, Carroll S. Nature 1997;388:639–648.

axis and within developing limbs suggest that the ancestral colinearity of *Hox* expression was maintained as these genes assumed novel roles in limb development.

Another novel feature of posterior *Hox* gene regulation is evident in the distal element of tetrapod limbs. The **autopod**, which consists of the hand (or foot) and digits, is unique to tetrapods. During autopod development, a late phase of *Hox* gene expression (phase III, see Chapter 3) appears that is not present in teleost fish (Fig. 6.8). This phase is characterized by a reversal in the anteroposterior order of *Hox* gene expression in the autopod compared with the earlier, more proximal expression domains in other developing limb elements. A single *cis*-regulatory element drives phase III expression of all of the posterior *HoxD* genes, indicating that several *HoxD* genes in the autopod may have been recruited simultaneously through the evolution of this element.

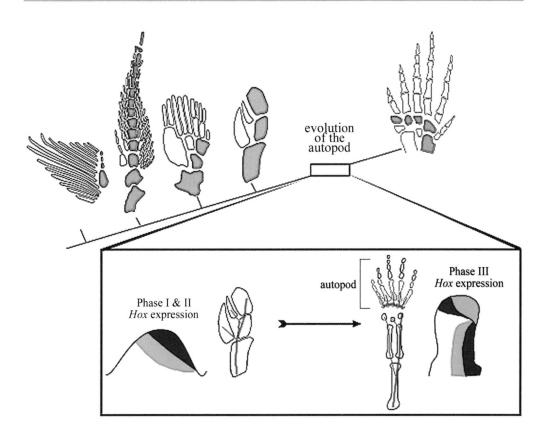

Figure 6.8
The origin of tetrapod digits

The evolution of the autopod is characterized by branching of digits from a bend in the limb axis and by the appearance of a third phase of *Hox* gene expression in the distal limb. A fin axis is present in the paired fins of bony fish (gray; first four fins on **left**). The distal tetrapod limb has a 90° bend in the equivalent limb axis (gray; **top right**). (**box**) Proximal elements of the vertebrate limb branch to the posterior from the limb axis, whereas elements of the autopod branch to the anterior (red). The reversal in branching direction is also reflected in a reversal in the nested domains of phase III *Hox* gene expression (light blue, *Hoxd11*; dark blue, *Hoxd13*).

Source: Modified from Shubin N, Tabin C, Carroll, S. Nature 1997;388:639–648.

Evolution of the neural crest and the vertebrate brain

A new embryonic cell population, the neural crest, evolved during the elaboration of the modern vertebrate body plan. The neural crest is derived from cells found at the interface of the lateral neural plate and epidermis along the dorsal side of the body. These cells migrate extensively within the body to form a variety of structures and cell types, including the facial skeleton, connective tissue in the head and neck, peripheral neurons and glia, and melanocytes. The appearance of the neural crest may have facilitated the evolution of larger body sizes and more complex behaviors, including social interactions and predation, in the vertebrate lineage.

Basal vertebrate lineages such as hagfish and lampreys, as well as early fossil vertebrates, appear to possess most of the neural crest derivatives, indicating that this cell population arose early in the vertebrate lineage. Other chordates, such as amphioxus and ascidians, do not have a true neural crest cell population. Nevertheless, comparisons of gene expression have suggested that members of the regulatory circuits that are required for neural crest induction in vertebrates are deployed in the neural tube in more basal chordates. Vertebrate neural crest cells express a suite of developmental regulatory genes, including members of the *Dlx, Msx, slug/snail,* and *Pax3/7* families. Homologs of these genes are expressed in the lateral neural plate of both amphioxus and ascidians. Thus the spatial expression domains of these genes appear to have been established in the lateral neural plate before neural crest cells evolved at the base of the vertebrate lineage. The elaboration of the vertebrate neural crest, with its diverse neuronal and non-neuronal cell types, may have its roots in a small ancestral population of cells in or near the chordate lateral neural plate.

In concert with the emergence and proliferation of neural crest cells, vertebrates evolved a complex, highly regionalized brain. The developing vertebrate brain is divided into forebrain, midbrain, and hindbrain. The most anterior region of the forebrain (the telencephalon) has clearly grown in size and complexity during the evolution of higher vertebrates. In mammals, a unique cortical region of the telencephalon (the cerebral cortex) has expanded to account for the majority of the brain and has taken over most sensory and motor functions. The cerebral cortex of a human brain, for example, includes as many as 100 billion neurons, organized into stratified layers, with countless synaptic interconnections.

The subdivision of the vertebrate brain into forebrain, midbrain, and hindbrain preceded the development of the telencephalon and the mammalian neocortex. This initial regionalization of the vertebrate brain may reflect an ancestral regionalization of the anterior chordate central nervous system (CNS), as indicated by patterns of regulatory gene expression (Fig. 6.9). In ascidians and amphioxus, the *Otx* gene is deployed in the most anterior regions of the CNS, similar to the expression of *Otx* genes in the vertebrate forebrain and midbrain. In both ascidians and vertebrates, the region immediately posterior to the *Otx*

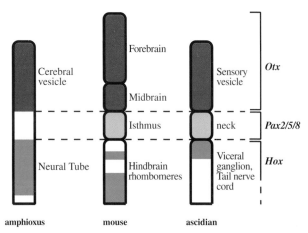

Figure 6.9
Elaboration of the vertebrate brain

The anterior central nervous system greatly expanded and acquired more complex organization during vertebrate evolution. The developmental regionalization of the vertebrate brain is marked by the expression of *Otx, Pax-2/5/8,* and *Hox* genes. Similar patterns of deployment in basal chordates may indicate homologous regions within the developing brains of amphioxus, mouse, and ascidian embryos.

Source: Redrawn from Williams NA, Holland PWH. Brain Behav Evol 1998; 52:177–185.

domain is marked by expression of *Pax2/5/8* genes. The rhombomeres of the vertebrate hindbrain are marked by expression of the *Hox* genes (see Chapter 3). In much the same way, the *Hox* genes of amphioxus and ascidians are deployed in nested domains posterior to the *Otx* domain. The similarities in regulatory gene expression in these chordates indicates that their last common ancestor likely had a regionalized anterior CNS, with discrete domains of *Otx, Pax2/5/8*, and *Hox* gene expression.

The increased complexity of vertebrate neural development, including both the elaboration of the forebrain/telencephalon and the appearance of a neural crest cell population, may be correlated with expansion of the vertebrate genome through large-scale duplication events (see Chapter 4). Perhaps the additional developmental regulatory genes that were created by gene duplication evolved novel roles in controlling the differentiation of new cell types and the more complex organization of the vertebrate CNS. In vertebrates, gene duplication and divergence may have provided the *potential* for more complex morphologies and, in the case of the brain, a greater range of complex behaviors.

Evolution of the notochord

The presence of a notochord unites the chordates and is the distinguishing feature that inspired their name. The notochord is a stiff, axial rod of cells that represents the functional precursor of the vertebral column in basal chordates. It acts as an organizer for the early development of the CNS and adjacent axial mesoderm. Efforts to determine the origin of the notochord have often focused on the urochordates because of their relatively simple body plan and their basal position within the chordates. The larval ascidian tail consists of the notochord, muscle cells, and a dorsal nervous system. At metamorphosis, ascidian larvae lose their entire tail (including the notochord) and become sessile, filter-feeding adults.

The evolution of the notochord involved the co-option of an ancestral regulatory gene, *Brachyury* (*T*), a member of the T-box class of transcription factors (Fig. 6.10). In vertebrates, *T* is expressed in developing notochord cells and other mesodermal derivatives and is required for notochord differentiation. Similarly, the ascidian *T* gene is expressed in cells that form the notochord and is sufficient to confer notochord fate. In invertebrates that lack a notochord, homologs of the *T* gene are expressed in posterior mesoderm. This mesodermal expression may reflect the ancestral function of *T*—that is, the role it played before the evolution of the notochord. This gene began to play a new role in notochord development after the chordate and hemichordate/echinoderm lineages became separated, as the hemichordate *T* gene is not expressed in the stomochord (a possible precursor of the notochord).

EVOLUTION OF RADICAL BODY PLAN CHANGES

Morphological novelty also encompasses the evolution of novel body organizations. Radical reorganization of body plans can include both the appearance of new structures and the loss of ancestral characters. The examples we discuss here have lost one or more key morphological features characteristic of their clade. As with the other types of novelties discussed in this chapter, these morphological changes are correlated with changes in genetic regulatory circuits.

Loss of the ascidian tail: a chordate lacking chordate features

Although the notochord is a defining feature of the chordates, several ascidian species have an abbreviated larval stage that has lost the notochord along with the entire larval tail. Of

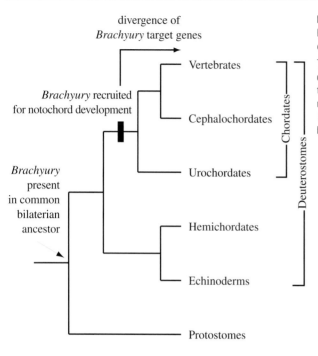

divergence of
Brachyury target genes

Brachyury recruited
for notochord development

Brachyury
present
in common
bilaterian
ancestor

Vertebrates

Cephalochordates

Urochordates

Hemichordates

Echinoderms

Protostomes

Chordates

Deuterostomes

Figure 6.10

Recruitment of the *Brachyury* gene during evolution of the notochord

The *Brachyury* (*T*) gene predates the origin of the chordate notochord. During the early evolution of the chordates, expression of the *T* gene was recruited to pattern the notochord. Targets of the *T* gene have evolved as the chordate lineages have diverged.

the more than 3000 ascidian species known, about a dozen are tailless. Taillessness is associated with direct development, including a rapid progression to the sessile adult form.

In some cases, very closely related species differ dramatically in the extent of their tail development. For example, the urodele (tailed) larvae of *Molgula oculata* develop a notochord, a spinal cord, and striated muscle cells. In contrast, another *Molgula* species (*M. occulta*) has an anural (tailless) larval stage that lacks these characteristic chordate features. When these species are interbred in the laboratory, the crosses yield a short-tailed hybrid, complete with notochord (Fig. 6.11). Thus the tailless phenotype of *M. occulta* can be rescued by the *M. oculata* genome, suggesting that *M. occulta* may have lost some genetic function that results in its tailless larval form.

The search for genes that are essential for *M. oculata* tail development, but that are downregulated in *M. occulta*, has led to the identification of the *Manx* and *Bobcat* genes. The *Manx* gene is a zinc-finger transcription factor, and *Bobcat* encodes an RNA helicase. The temporal and spatial expression of these genes appears to be jointly regulated in the embryonic cells that will give rise to the *M. oculata* larval tail (including the notochord). Both genes are similarly deployed in *M. oculata* x *M. occulta* hybrids, and inhibition of *Manx* or *Bobcat* gene activity generates a tailless hybrid embryo. The requirement for *Manx* and *Bobcat* gene function indicates that these genes are part of the developmental program for tail formation.

The reduced expression of the *Manx* and *Bobcat* genes in the tailless species *M. occulta* suggests that the tail regulatory hierarchy is terminated early in development. During the evolution of this tailless form, the expression or function of an upstream regulator of *Manx* and

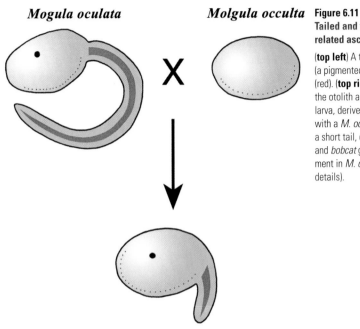

Mogula oculata   *Molgula occulta*

Figure 6.11
Tailed and tailless forms of two closely related ascidian species

(**top left**) A tailed *M. oculata* larva has an otolith (a pigmented organ in the head) and a notochord (red). (**top right**) A tailless *M. occulta* larva lacks the otolith and notochord. (**bottom**) A hybrid larva, derived from a *M. occulta* egg fertilized with a *M. oculata* sperm, displays an otolith and a short tail, complete with notochord. The *Manx* and *bobcat* genes are essential for tail development in *M. oculata* and in the hybrid (see text for details).

Bobcat may have been changed that led to the loss of expression of these genes. Alternatively, the evolutionary switch to taillessness may have acted directly within a shared *cis*-regulatory region of *Manx* and *Bobcat* themselves. The altered expression of a single regulatory gene required for tail development may be sufficient to evolve a novel larval form.

Limbless tetrapods, or how the snake lost its legs

The evolution of snakes, with their characteristic elongated bodies and reduced limbs, represents the adaptation of the tetrapod body plan to a novel form. Body elongation and limblessness are associated with burrowing and have evolved independently many times within the reptiles, including once at the base of the snake lineage. Among modern snakes, the python lineage retains some features of the tetrapod hindlimb, including a pelvic girdle and rudimentary hindlimbs, whereas more highly derived snakes are completely limbless. Comparisons between snakes and other tetrapods have revealed key developmental genetic differences in axial patterning and limb development that correlate with the evolutionary transition to the limbless snake body plan.

The concomitant loss of snake forelimbs and expansion of thoracic axial identities suggest that a common genetic mechanism may underlie these evolutionary transitions. In Chapter 5, we discussed the correlation between the python axial skeleton identities and the anterior expansion of the expression domains of *Hoxc6* and *Hoxc8*. In other tetrapods, the axial position of the forelimb is determined by the anterior boundary of *Hoxc6* expression in lateral plate mesoderm at the cervical–thoracic transition (see Fig. 5.6). Thus the extension of the expression pattern of python *Hoxc6* and *Hoxc8* up to the cranial region may disrupt the *Hox*-regulated positioning of the forelimb. Without the proper axial patterning cues, this limb bud never forms. The correlation between trunk elongation and limb loss in several other vertebrate taxa may be explained by similar changes in axial patterning caused by altered *Hox* expression patterns.

The fossil record and developmental studies indicate the reduction and loss of the snake hindlimb occurred via a separate series of evolutionary events. The initial development of the vestigial python hindlimb appears normal, indicating that the axial specification cues and early induction of python hindlimb development remain intact. As in other tetrapods, the posterior boundaries of *Hoxc6* and *Hoxc8* expression mark the python thoracic–lumbar transition at the position of the rudimentary pelvic girdle, and a limb bud does form. Ultimately, however, development of the hindlimb is arrested and no outgrowth is observed.

The python hindlimb bud lacks some of the characteristic features of other developing tetrapod limbs, such as an apical ectodermal ridge (AER) and a zone of polarizing activity (ZPA). Nevertheless, experimental manipulations have shown that the python limb bud mesenchyme retains the capacity to induce an AER and a ZPA; indeed, it can form a complete limb in response to an inductive signal from an exogenous AER. Thus much of the developmental program of limb formation is present in the python limb bud, but at least one essential element must be missing. The termination of python hindlimb development suggests that this field fails to initiate or maintain the signaling events that are required for AER formation and hindlimb outgrowth.

The loss of a structure is not necessarily accompanied by the loss of the genes—or even genetic regulatory circuitry—required to build that structure. The condition of limblessness can be reversed, as evidenced both by rare individuals in wild populations and by stable evolutionary lineages. For example, rare natural occurrences of cetaceans (whales and dolphins) with well-developed hindlimbs have been reported. Also, the snake fossil record shows that a limbless ancestor may have given rise to a group of snakes that had substantial limbs (Fig. 6.12). For example, the fairly well-developed hindlimbs of the snake fossils *Pachyrhachis problematicus* and *Hassiophis terrasanctus* may represent a reversal from a limbless state. Based on these cases of reversion to the ancestral limbed condition, it appears that the limb developmental program may still be present in limbless animals. As the genes required for limb patterning have other developmental functions, they are retained in the genomes of limbless taxa. More surprising, however, is that an otherwise cryptic regulatory circuitry required for limb development is retained in a limbless tetrapod.

Evolution of the echinoderm body plan

The defining features of the Bilateria, including the evolution of bilateral symmetry with distinct rostrocaudal and D/V axes, arose before the radiation of most animal phyla. Although the echinoderms (sea stars, sea urchins, sea cucumbers, brittle stars, and crinoids) are a basal deuterostome lineage derived from a bilaterian ancestor, echinoderm adults have a radially symmetric body organization. Whereas echinoderm larvae exhibit a more typical bilateral form, adult echinoderms are radically rearranged, making it difficult to identify any vestiges of bilateral symmetry (Fig. 6.13). The fivefold symmetry of the adult nervous system, connected by a central nerve ring, has led to the proposal that each echinoderm "arm" represents a duplication of the bilaterian A/P axis. Alternatively, the distribution of mesodermal tissue in the adult may indicate that the orthogonal adult oral–aboral

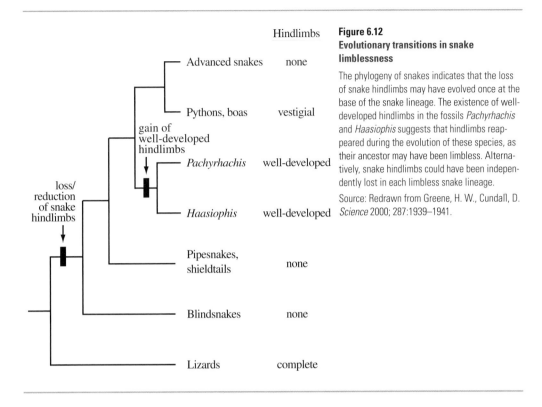

Figure 6.12
Evolutionary transitions in snake limblessness

The phylogeny of snakes indicates that the loss of snake hindlimbs may have evolved once at the base of the snake lineage. The existence of well-developed hindlimbs in the fossils *Pachyrhachis* and *Haasiophis* suggests that hindlimbs reappeared during the evolution of these species, as their ancestor may have been limbless. Alternatively, snake hindlimbs could have been independently lost in each limbless snake lineage.

Source: Redrawn from Greene, H. W., Cundall, D. *Science* 2000; 287:1939–1941.

axis is derived from the larval A/P axis. The origin of the unique and highly derived body plan of adult echinoderms remains a puzzle.

The development of other structural novelties of echinoderms, including the water vascular system and the calcitic endoskeleton, appears to involve regulatory genes shared with other phyla. These genes have been recruited for deployment in echinoderm-specific tissues in the developing juvenile sea urchin. *Dll* and *Otx* are expressed in the tube feet (podia), which are extensions of the water vascular system. *Hox3* is expressed in the nascent tooth sacs, which give rise to the echinoderm mouthparts. The pattern of *en* expression in the ectoderm reflects the organization of the underlying endoskeleton. In short, these regulatory genes appear to have been co-opted for new patterning roles associated with the evolution of novel echinoderm structural elements.

REGULATORY EVOLUTION AND THE ORIGIN OF NOVELTIES

The recurring theme among the diverse examples of evolutionary novelties described in this chapter is the creative role played by evolutionary changes in gene regulation. The evolution of new regulatory linkages—between signaling pathways and target genes, transcriptional regulators and structural genes, and so on—has created new regulatory circuits

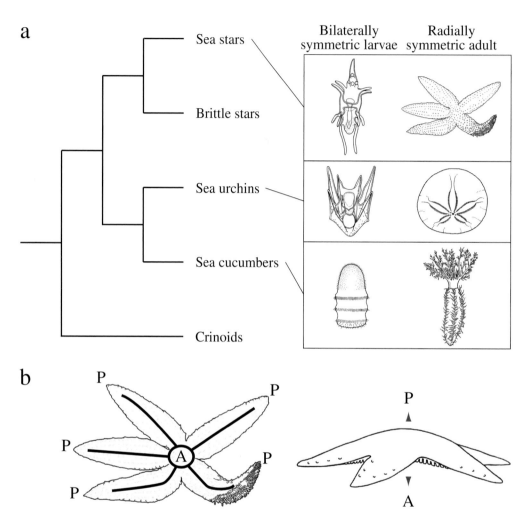

Figure 6.13
Evolution of the radially symmetric echinoderm body plan

(a) Echinoderm lineages have bilaterally symmetric larvae and radially symmetric adults. (b) Two hypotheses regarding the orientation of the ancestral bilaterian A/P axis in the adult echinoderm body plan: (**left**) multiplication of the bilaterian A/P axis may have created the radial nerve cords that extend down each echinoderm arm; (**right**) the ancestral bilaterian A/P axis may have evolved into the echinoderm oral–aboral axis.

that have shaped the development of myriad functionally important structures. These regulatory circuits also serve as the foundation of further diversification. The patterning of structures such as the tetrapod autopod and butterfly wing has diverged extensively throughout the radiation of the lineages in which these novelties first evolved.

The time scale over which new structures have evolved or body plans have diverged is considerable—on the order of many millions of years. One of the best-analyzed transitions, the evolution of tetrapod limbs from fish fins, is estimated to have transpired over the course of about 9 million years, for example. As we learn more about the architecture of the genetic regulatory differences between lineages, as well as the scope and time scale of anatomical evolution, we can better appreciate that changes in a potentially very large number of regulatory linkages accompany morphological evolution. In Chapter 7, we will examine how variation in genetic regulatory circuits and morphological characters arises and how this variation provides the raw material for the evolution of diverse animal forms.

SELECTED READINGS

General Reviews

Gerhart, J., Kirschner, M. Cells, embryos, and evolution. Malden, MA: Blackwell Science, 1997.

Müller, G. B., Wagner, G. P. Novelty in evolution: restructuring the concept. *Ann Rev Ecol Systematics* 1991; 22:229–256.

Raff, R. The shape of life. Chicago: University of Chicago Press, 1996.

Epithelial Appendages

Chuong, C. M., ed. Molecular basis of epithelial appendage morphogenesis. Austin, TX: R. G. Landes, 1998.

Crowe, R., Niswander, L. Disruption of scale development by Delta-1 misexpression. *Develop Biol* 1998; 195:70–74.

Godwin, A. R., Capecchi MR. *Hoxc13* mutant mice lack external hair. *Development* 1988; 12:11–20.

Widelitz, R. B., Jiang, T. X., Lu, J., Chuong, C. M. *beta-catenin* in epithelial morphogenesis: conversion of part of avian foot scales into feather buds with a mutated *beta-catenin*. *Develop Biol* 2000; 219:98–114.

Zou, H., Niswander, L. Requirement for BMP signaling in interdigital apoptosis and scale formation. *Science* 1996; 272:738–741.

Origin of the Insect Wing

Averof, M., Cohen, S. M. Evolutionary origin of insect wings from ancestral gills. *Nature* 1997; 385:627–630.

Jockusch, E. L., Nagy, L. M. Insect evolution: how did insect wings originate? *Curr Biol* 1997; 7:R358–R361.

Butterfly Scales and Eyespots

Galant, R., Skeath, J. B., Paddock, S., et al. Expression of an *achaete-scute* homolog during butterfly scale development reveals the homology of insect scales and sensory bristles. *Curr Biol* 1998; 8:807–813.

Keys, D. N., Lewis, D. L., Selegue, J. E., et al. Recruitment of a *Hedgehog* regulatory circuit in butterfly eyespot evolution. *Science* 1999; 283:532–534.

Vertebrate Limbs and Digits

Coates, M. The origin of vertebrate limbs. *Development* 1994; (suppl):169–180.

Herault, Y., Beckers, J., Gerard, M., Duboule, D. *Hox* gene expression in limbs: colinearity by opposite regulatory controls. *Develop Biol* 1999; 208:157–165.

Nelson, C. E., Morgan, B. A., Burke, A. C., et al. Analysis of Hox gene expression in the chick limb bud. *Development* 1996; 122:1449–1466.

Shubin, N., Tabin, C., Carroll, S. Fossils, genes and the evolution of animal limbs. *Nature* 1997; 388:639–648.

Sordino, P., van der Hoeven, F., Duboule, D. Hox gene expression in teleost fins and the origin of vertebrate digits. *Nature* 1995; 375:678–681.

Evolution of the Neural Crest, the Vertebrate Brain, and the Notochord

Holland, N. D., Panganiban, G., Henyey, E. L., Holland, L. Z. Sequence and developmental expression of AmphiDll, an amphioxus Distal-less gene transcribed in the ectoderm, epidermis and nervous system: insights into evolution of craniate forebrain and neural crest. *Development* 1996; 122:2911–2920.

Satoh, N., Jeffery, W. R. Chasing tails in ascidians: developmental insights into the origin and evolution of chordates. *Trends Genetics* 1995; 11:354–359.

Shimeld, S. M., Holland, P. W. H. Vertebrate innovations. *Proc Natl Acad Sci USA* 2000; 97: 4434–4437.

Wada, S., Saiga, H., Satoh, N., Holland, P. W. H. Tripartite organisation of the ancestral chordate brain and the antiquity of placodes: insights from ascidian *Pax-2/5/8, Hox* and *Otx* genes. *Development* 1998; 125:1113–1122.

Williams, N. A., Holland, P. W. H. (1998). Molecular evolution of the brain of chordates. *Brain Behav Evol* 1998; 52:177–185.

Zimmer, C. In search of vertebrate origins: beyond brain and bone. *Science* 2000; 287:1576–1579.

Loss of Notochord in Ascidians

Swalla, B. J., Jeffery, W. R. Requirement of the *Manx* gene for expression of chordate features in a tailless ascidian larva. *Science* 1996; 271:1205–1208.

Swalla, B. J., Just, M. A., Pederson, E. L., Jeffery, W. R. A multigene locus containing the *Manx* and *bobcat* genes is required for development of chordate features in the ascidian tadpole larva. *Development* 1999; 126:1643–1653.

Loss of Limbs in Snakes

Cohn, M. J., Tickle, C. Developmental basis of limblessness and axial patterning in snakes. *Nature* 1999; 399:474–479.

Greene, H. W., Cundall, D. Limbless tetrapods and snakes with legs. *Science* 2000; 287:1939–1941.

Echinoderm Body Plan

Arenas-Menas, C., Martinez, P., Cameron, R. A., Davidson, E. H. Expression of the Hox gene complex in the indirect development of a sea urchin. *Proc Natl Acad Sci USA* 1998; 95:13062–13067.

Lowe, C. J., Wray, G. A. Radical alterations in the roles of homeobox genes during echinoderm evolution. *Nature* 1997; 389:718–721.

Peterson, K. J., Arenas-Mena, C., Davidson, E. H. The A/P axis in echinoderm ontogeny and evolution: evidence from fossils and molecules. *Evol Develop* 2000; 2:93–101.

CHAPTER 7

From DNA to Diversity: The Primacy of Regulatory Evolution

The assembly and expansion of the genetic toolkit in early bilaterians and its roughly fourfold expansion in early vertebrates were two intervals in which gene duplication and divergence played major roles in morphological evolution. However, the diversification of protostomes, most deuterostomes, and the vertebrates (after genome expansion) occurred around ancient and largely equivalent toolkits of major developmental genes. These observations, which were illustrated by case studies in Chapters 5 and 6, lead us to conclude that changes in developmental gene regulation are the predominant genetic mechanisms underlying large-scale morphological evolution. This idea prompts a question: Are smaller-scale variations and differences in morphology within or between species also regulatory in nature? And, given that variation is the raw material of evolution, are the genetic mechanisms underlying differences at or below the species level sufficient to account for the larger-scale patterns of morphological evolution?

In the final chapter of this book, we consider why regulatory evolution is the creative force underlying morphological diversity across the evolutionary spectrum, from variation within species to body plans. The link between evolution at the DNA level and phenotypic diversity involves the *cis*-regulatory elements acting as units of evolutionary change. Here, we take an in-depth look at *cis*-regulatory elements and molecular models for their evolution. We examine a few experimental approaches that have begun to reveal the direct connection between changes in regulatory DNA and morphological variation within species and the response to natural or artificial selection. Finally, we address the sufficiency of regulatory evolution to account for the larger-scale patterns of morphological evolution.

WHY IS REGULATORY EVOLUTION A PRIMARY FORCE IN MORPHOLOGICAL EVOLUTION?

The case studies in Chapters 5 and 6 illustrated that selected changes in gene regulation in one part of the developing animal,

"If it were possible to take judiciously chosen structural genes and put them together in the right relationship with regulatory elements, it should be possible to make any primate, with some small variations, out of human genes. . . . Likewise it should be possible to make any crustacean out of the genes of higher Crustacea."
—Emile Zuckerkandl (1976)

independent of other parts, underlie the morphological diversification of serially homologous structures and the origin of novelties. Regulatory evolution is the enabling genetic mechanism for the modular organization and diversity of large bilaterians and for the emergence of new morphological characters.

The argument made for the central importance of regulatory change in morphological evolution is not a new one. Indeed, the creative potential of regulatory change and the comparatively greater constraints on protein evolution were recognized early in the history of molecular biology. What is different now is that considerable empirical evidence exists to demonstrate that changes in the regulation of genes that affect morphology are implicated much more frequently in the evolution of diversity than are new genes or functional changes in protein sequences. Furthermore, our current understanding of genetic regulatory hierarchies, networks, and circuits in development reveals why this is so.

Specifically, regulatory evolution is powerful because of the following characteristics:

1. *Regulatory evolution enables pleiotropy of toolkit genes.* The same transcription factor usually controls different target genes in different tissues at different stages of development. The biochemical activities of a given protein in terms of DNA binding are the same throughout development. What evolves are specific regulatory sequences in target genes that enable them to respond in a context-specific fashion.

2. *Regulatory evolution enables developmental modularity.* For anatomical structures to become different from serial homologs within an animal or from homologs in other animals, changes must evolve in the regulatory hierarchies that operate during the development of these structures. Due to their modular organization, changes in *cis*-regulatory elements allow changes in gene expression to occur in one structure independently of another (this independence is also referred to as *dissociation*). In metazoans, the great success (in terms of species diversity and adaptation to terrestrial, aquatic, and aerial environments) of highly modular body plans such as the arthropods, vertebrates, and annelids suggests that the modularity of developmental regulatory mechanisms at both the anatomical (for example, fields) and molecular (*cis*-regulatory elements) levels have facilitated this diversity.

3. *Regulatory evolution is a rich and continuous source of variation.* The *cis*-regulatory regions may occupy vast stretches of DNA and contain many independent functional elements. Changes in regulatory sequences within individual elements may subtly affect the level, timing, or spatial pattern of gene expression, and may do so very selectively in terms of the tissues and stages of development involved. Without any change in protein sequences, these quantitative, temporal, and spatial changes in the deployment of regulatory genes may affect the level, timing, and spatial expression of other developmental genes. Regulatory changes are often cryptic and have little effect on morphology, but they nevertheless create genetic variations with the potential to produce the morphological variation that is the raw material of evolution. Regulatory DNA is a rich and continuous source of potential genetic, developmental, and phenotypic variation, and thus evolutionary change.

4. *Regulatory evolution creates novelty.* The exploration of new morphologies is facilitated by new combinations of gene expression that can arise without changes in protein function.

THE FUNCTION AND EVOLUTION OF *CIS*-REGULATORY DNA

Functions of cis-*regulatory elements*

Several examples of *cis*-regulatory elements and *cis*-regulatory regions were presented in Chapter 3 in the context of understanding the genetic regulatory networks that orchestrate pattern formation. To better understand the role played by *cis*-regulatory DNA in evolution, it is important to appreciate several general features of *cis*-regulatory systems.

First, most elements are regulated by a minimum of four to six different transcription factors of various structural types. Transcription factors virtually never act alone, so the output of individual *cis*-regulatory elements is determined by the integration of diverse and multiple inputs.

Second, the spatial relationship of binding sites for transcription factors within a *cis*-regulatory element can be of utmost functional importance. Within the few hundred base-pairs of a typical element, the proximity of binding sites can determine whether transcription factors interact cooperatively. For example, multiple lower-affinity sites may have a greater effect on transcription through cooperative interactions than a single high-affinity site. Most metazoan transcription factors bind a family of DNA sequences with some degree of degeneracy in the sequences recognized. The "flexibility" of transcription factor binding opens up greater opportunities for cooperative interactions.

Third, repression—not activation—is generally the ground state of gene expression. That is, the chromatin that packages DNA has generally repressive effects on transcription. The *cis*-regulatory elements represent sites where protein complexes are assembled that function to overcome this repression.

Fourth, spatial boundaries of gene expression in a cellular field are usually set by both positive and negative inputs into *cis*-regulatory elements. Loss of positive inputs (in *trans*) or of their binding sites (in *cis*) eliminates or contracts the spatial domains of gene expression. Conversely, loss of repressors or their sites expands spatial domains. The sites where activators and repressors bind may overlap; in those cases, competition for sites may determine gene activity. In other cases, they may not overlap and the activators and repressors modulate the output of the element through short-range (fewer than 100 bp) or long-range (more than 100 bp) effects on transcriptional functions.

Finally, expression of terminal differentiation genes, such as those encoding structural proteins that carry out functions specific to certain cell types, are often regulated by *cis*-elements that respond largely or entirely to positive transcriptional activators. Hence, genes found progressively farther downstream from the regulators that establish spatial domains in the developing embryo may not receive negative inputs to prevent their expression in inappropriate positions or cell types.

Evolution of cis-*regulatory elements*

Evolutionary changes in a given gene's expression may arise through a variety of mechanisms. That is, they may arise directly from alterations in the *cis*-regulatory DNA of the gene or they can occur indirectly, through changes in the deployment or activity of upstream transcription factors that regulate the gene (which ultimately reflect changes in the *cis*-regulatory elements of genes encoding these transcription factors or their regulators). In the remainder of this chapter, we focus on the developmental and evolutionary consequences of changes in *cis*-regulatory elements.

Cis-regulatory DNA function may evolve through any of several molecular mechanisms. Here, we distinguish between two major potential sources of *cis*-regulatory DNA evolution:

- The de novo evolution of *cis*-regulatory elements through changes in nonfunctional DNA (Fig. 7.1a)
- The evolution of *cis*-regulatory elements from preexisting functional elements (Fig. 7.1b–d)

Within the latter category, we distinguish between three modes of sequence evolution:

- Duplications and DNA rearrangements involving existing functional elements that create additional copies of elements or new elements (see Fig. 7.1b)
- Modifications of existing *cis*-regulatory elements—for example, through the gain or loss of binding sites for positive and/or negative regulators (see Fig. 7.1c)
- The special case of co-option of an existing element that expands its developmental function (see Fig. 7.1d)

First, we examine, from a theoretical viewpoint, the factors affecting the probabilities of these different mechanisms of *cis*-regulatory DNA evolution. Next, we analyze some illuminating case studies of *cis*-regulatory element evolution.

De novo evolution of* cis-*regulatory elements from preexisting nonfunctional DNA sequences. The vast majority of animal genomes are composed of noncoding DNA. Consequently, extensive stretches of DNA sequences exist that could *potentially* function to regulate the expression of adjacent genes. Much of this DNA consists of various lengths of repetitive elements, some of which may have architectural functions (for example, the chromosome centromere), but most of which are believed to be nonessential and not involved in the regulation of specific genes. In addition, an abundant amount of single-copy DNA appears between coding regions.

In principle, such unconstrained sequences could evolve into functional *cis*-regulatory elements. The key question is, How "difficult" is it for random DNA sequences to evolve into a new regulatory element? To answer this question, we must consider two major factors. The first factor is the probability that one or more binding sites evolve for individual specific DNA-binding proteins. The second consideration is the minimal input required for *cis*-element function; more specifically, a novel element requires a level of function upon which positive selection can act.

The probability that binding sites for transcription factors will evolve depends on the properties of DNA-binding proteins. Individual DNA-binding proteins typically recognize a 5 to 9 bp core sequence, usually with some flexibility (that is, degeneracy) in the identity of bases found at particular positions. Assuming an equal frequency of all four bases in a genome, we can calculate the frequency with which a given length site will appear in a genome (Table 7.1). For example, a specific 6 bp sequence will occur, on average, once every 4096 bases. We can also relate these frequencies to genome size and to the approximate number of genes in various animal genomes (Table 7.1). For example, in a genome the size of that of *D. melanogaster* (approximately 1×10^8 bp) or of humans (approximately 3×10^9 bp), a given 6 bp sequence should occur roughly 25,000 and 750,000 times, respectively. Based on the number of genes in these species, these calculations demonstrate that

a

De novo evolution from non-functional sequences

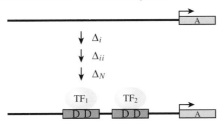

b

Duplications and DNA rearrangements of existing cis-regulatory elements

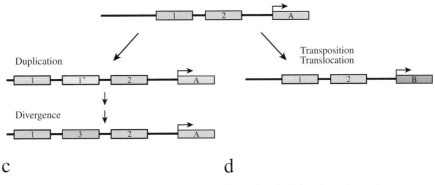

c

Modification of existing cis-regulatory elements

d

Co-option of existing cis-regulatory elements

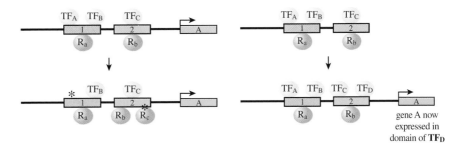

gene A now
expressed in
domain of **TF$_D$**

Figure 7.1
Molecular mechanisms of *cis*-regulatory DNA evolution

Four modes of *cis*-regulatory element evolution are depicted. (**a**) De novo evolution of an element from nonfunctional sequences could occur through mutations (Δ_i, Δ_{ii}, and so on) that create binding sites for transcription factors (TF) in proximity to a transcription unit. (**b**) New elements may evolve by duplications of genes or of *cis*-regulatory elements, followed by sequence divergence (**left**), or DNA rearrangements that move elements to new sites in the genome (**right**). (**c**) Modifications of existing *cis*-regulatory elements can occur through the gain, loss, or change in affinity of sites for activators (TF) or repressors (R). (**d**) A special case of *cis*-regulatory element modification is co-option of an existing element to function in a new domain through the gain of sites for specific factors (TF$_D$).

TABLE 7.1 *The probable frequencies of various length sites for DNA-binding proteins in animal genomes*

Binding Site Length (bp)	Frequency of site (1/# of base pairs)	Approx. Number of Sites in Genome; per Gene	
		Drosophila[a]	Homo sapiens[b]
5	1024	100,000; 6.6	3,000,000; 30
6	4096	25,000; 1.6	750,000; 7.5
7	16,354	6,250; 0.4	187,500; 1.9
8	65,536	1562; 0.1	46,875; 0.5
9	262,144	390; 0.03	11,718; 0.1
10	1,048,576	97; 0.007	2,929; 0.03
12	16,777,216	6; 0.0004	183; 0.002

[a] Assumes approx 1×10^8 bp genome; 15,000 genes
[b] Assumes approx 3×10^9 bp genome; 100,000 genes

single binding sites for proteins with a six-base core recognition sequence will occur, on average, at least once per gene.

These frequency calculations apply only to a single genome at a given time. Obviously, species are composed of potentially very large populations with characteristic generation times. Over the course of evolutionary time, the steady occurrence of random mutations will increase the probability that a given binding site will arise in a given genomic location at some time in a population.

Next, we must ask whether the appearance of any binding site is functionally significant. Answering this question requires that we consider the requirements for *cis*-regulatory element function, the diversity of the transcription factor repertoire, and the level of gene expression upon which selection can act. *Cis*-regulatory element function typically depends on a host of binding sites for diverse regulatory proteins acting in some proximity to one another. Gene activation must overcome the generally repressive effects of chromatin through the recruitment of co-activator proteins by transcription factors. In most elements, single binding sites or single transcription factors are not sufficient to confer significant transcriptional activation. Likewise, gene activation typically involves the synergistic effects of two or more activators bound to nearby sites. If these requirements need to be met to achieve the levels of gene activation upon which selection can act, then our calculations in Table 7.1 must be extended to account for the probability of two or more sites arising within an interval of the typical length (about 200 bp) of a *cis*-regulatory element (Table 7.2).

For example, target gene activation by several Hox proteins involves interactions with the Exd homeodomain protein bound to an adjacent binding site. A typical Hox–Exd composite site contains a critical 10 bp core sequence. A specific 10 bp sequence will occur at random approximately once every 1 million bp (see Table 7.1). If DNA is measured in 200 bp intervals, such a rare site would appear at random only once per any 5000 intervals (see Table 7.2). In comparison, combinations of shorter sequences would arise more frequently. For example, two specific 5 bp sequences are 200 times more likely to appear in a 200 bp interval than they are in a single 10 bp site (see Table 7.2).

TABLE 7.2 *Frequency of one or more various length sites of DNA-binding proteins in 200 bp interval of DNA sequence*

Binding Site Length (bp)	Approximate Frequency of *n* Sites per 200 bp[a]			
	n = 1	2	3	4
5	$1/5$	$1/25$	$1/125$	$1/625$
6	$1/20$	$1/400$	$1/8000$	$1/160,000$
7	$1/80$	$1/6400$	$1/512,000$	$1/40,960,000$
8	$1/320$	$1/162,400$	$1/32,768,000$	$<10^{-10}$
9	$1/1280$	$1/1,638,400$	$<10^{-9}$	$<10^{-12}$
10	$1/5120$	$1/26,214,400$	$<10^{-11}$	$<10^{-14}$

[a] Rounded to simplify Table.

Experiments have shown that transcription factors can often interact synergistically with a variety of other transcription factors. This finding implies that for a new binding site to function within an evolving element, the only requirement may be that other transcription factors bind nearby. The identity or type of transcription factor may not be critical. For example, classic studies of steroid receptor-binding sites have demonstrated that a wide array of transcription factors can interact synergistically with receptor-DNA complexes to induce gene activation. Thus a steroid receptor-binding site (or, in principle, a site for any other DNA-binding activator protein) that arises at random could impart activity upon a nearby gene if nearby DNA sequences are also bound by other transcription factors.

To analyze the probability that nearby binding sites will be occupied, we must have an understanding of the transcription factor repertoire in a typical animal cell. Such information is just emerging from genome sequencing and genome-wide screens of gene expression. As a result, we are beginning to get an idea of the size and complexity of the entire transcription factor toolkit.

In *C. elegans,* for example, of the roughly 19,000 protein coding genes in the genome, a few more than 400 are predicted to encode DNA-binding transcription factors. To calculate the probability that one or more of these transcription factors will bind to sites near each other or the site of another transcription factor, we must make some inferences about the diversity of transcription factors expressed in any given cell. The 400 or so transcription factors in *C. elegans* are expressed in a total of about 30 cell types. Although some fraction of these transcription factors may be expressed in all cells, the distribution of most likely remains restricted to a few cell types. For discussion purposes, we'll assume that each transcription factor acts in 5 cell types on average. We can extrapolate that, if each of these 400 different factors is expressed in 5 cell types, then the average cell contains 400 × $5/30$ or roughly 60 different transcription factors.

Assuming an average 6 bp recognition sequence, for any 20 transcription factors with nonidentical binding specificities, at least one site probably exists for one of the 20 transcription factors every 200 bp of DNA sequence. With 60 factors per cell, then at least 3 sites for the average set of transcription factors would exist in every 200 bp. These calculations suggest that for any randomly evolved site, a few other potential sites for different transcription factors are likely to exist nearby. A new site appearing in this context may be sufficient to affect gene activity.

The theoretical considerations discussed in this example largely focus on the probabilities surrounding the evolution of 6 bp sites. In reality, many transcriptional regulators, such as homeodomain proteins, recognize shorter core sequences and exhibit a range of measurable affinities for families of related sequences. Homeodomains, for example, have evolved to be able to interact with many different DNA sequences. An important implication of this flexibility of homeodomains (and potentially other regulatory proteins) is that they can bind to shorter, more abundant, lower-affinity sites. Such sites may be functional intermediates in the course of *cis*-regulatory DNA evolution upon which positive selection could act.

Evolution of* cis-*regulatory elements from existing elements.   General chromatin repression and the calculations given in the previous example suggest that limitations and uncertainties surround the probability of de novo element evolution. Surmounting these limitations through the accumulation of single, initially neutral base changes is not the only means of evolving new functional elements. Indeed, one way to evolve new elements more readily might be to derive them from preexisting functional elements. The three other *cis*-regulatory element evolution scenarios depicted in Figs. 7.1(b–d) are all variations on this theme, albeit with important distinctions that are delineated below.

Duplications and DNA rearrangements   Perhaps the simplest way that genomes can "experiment" with the expression of individual genes is by shuffling *cis*-regulatory DNA in the genome (see Fig. 7.1b). Tandem duplications expand the number of *cis*-regulatory elements, and recombination, transposition (for example, of mobile elements), inversions, and translocations all create new juxtapositions of DNA sequences. If functional *cis*-regulatory DNA (including all or part of an element) becomes newly juxtaposed to another *cis*-regulatory element or transcription unit, the rearrangement may affect gene expression. In addition, new combinations of regulatory sequences may be created that have properties differing from those of either the new or resident sequence.

Because regulatory DNA is not constrained to maintain any reading frame or spacing, *cis*-regulatory DNA is more likely than coding regions to tolerate and thus accumulate partial duplications and insertions that expand the overall element. Local duplications of elements or entire genes can create functional redundancy that may be retained by positive selection because of the duplication's effects on gene expression levels. Expanded elements with duplicated sites may be more tolerant both to substitutions that can cause functional divergence and to DNA rearrangements that separate or subdivide parts of an element. Acting in this manner, existing *cis*-regulatory DNA elements could give rise to new elements much more readily than could the gradual accumulation of single base changes.

Modification of existing cis-regulatory elements   Any *cis*-regulatory element is subject to the constant forces of mutation and selection. Changes in nucleotide sequence will be neutral if they meet any of the following criteria:

- They fall outside of any site recognized by an essential regulator of an element.
- They do not alter the affinity of a critical binding-site.
- The change in a binding site has negligible effects on the overall function of the element (which may be due to other compensatory changes that occur).

Changes may not be neutral if they affect the timing, level, or spatial domain of gene expression controlled by the element. Changes in the affinity or the gain or loss of activator sites and repressor sites can subtly or more dramatically affect *cis*-element function.

For *cis*-regulatory elements that regulate gene expression patterns within a particular field, the domain of expression controlled by individual elements can expand or contract simply by a net decrease or increase in activator sites or a net decrease or increase in repressor sites (see Fig. 7.1c). For example, for a regulatory element that is controlled by a signal or transcription factor whose concentration is graded, gene expression could broaden or narrow, respectively, if a gain/reduction in the number or a change in the affinity of activator sites occurs. Alternatively, gene expression could be excluded from all or part of a field through the evolution of sites for a spatially localized repressor.

Co-option of an existing element and expansion of its developmental function   The evolution of a new *cis*-regulatory function could also arise through modification of an existing element (Fig. 7.1d). We consider this possibility here as a special case of modification of existing *cis*-regulatory elements, which may also subsequently involve mechanisms described for duplications and DNA rearrangements.

Given a functional element, if new sites for additional factors evolve, these new sites—acting in conjunction with functional sites bound by other transcription factors—might be able to co-opt an existing element such that it also functions in a new context (see Fig. 7.1d). Such a bifunctional or multifunctional element could then control gene expression in two or more tissues or developmental stages.

There are no mechanistic grounds to preclude the evolution of multifunctional elements. We can postulate, however, that functional pressures might potentially lead to additional evolutionary changes in multifunctional elements. Multifunctional elements lack the advantage of two or more modular elements. The necessity to "fine-tune" gene expression in two or more contexts may favor the subdivision of an element into multiple independent elements. The most facile way for this subdivision to occur is for additional modifications to accumulate, such as internal duplications and the addition of factor binding sites that enhance the functional robustness of the element and make its subdivision tolerable. A subdivision could arise through DNA rearrangements, including the interposition of sequences that functionally isolate *cis*-regulatory domains.

The evolution of gene repression versus gene activation

From these various mechanistic scenarios, we can see that the modification of gene expression patterns within a field can be readily explained by the accumulation of single base changes in *cis*-regulatory DNA or through DNA rearrangements. A corollary of these observations may be especially pertinent to the trends in body plan diversification described in this book. That is, the selective repression of part of an existing gene expression pattern may be more readily evolved at the *cis*-regulatory level than are novel tissue-specific patterns of gene expression.

We have emphasized the trend in vertebrate, arthropod, and annelid evolution marked by the increasing diversity of serially homologous structures. The diversification of somites, segments, appendages, and other parts often entails the selective repression of developmental regulatory networks in one homolog versus another (for example, limb and wing repression in the insect abdomen). If the evolution of single repressor binding sites in *cis*-regulatory elements is sufficient to modify or eliminate gene expression in a field, it would appear to involve fewer binding sites and transcription factors than is required for the evolution of new patterns of gene expression. If, indeed, gene repression is "easier" to evolve than gene activation is, then the morphological diversity that follows from selective repression of gene expression may be one of the most important evolutionary correlates of the logic of metazoan *cis*-regulatory element function.

Case studies in cis-regulatory evolution

The identification of specific cases representing any of the four modes of *cis*-regulatory element evolution is very challenging. While the preceding theoretical treatment outlines most of the conceivable possibilities, the analysis of the evolution of *cis*-regulatory elements of developmental genes remains in its early stages. This lag is partly explained by the facts that knowledge of *cis*-regulatory element function in development is far from complete and that it is difficult to identify all of the diverse inputs into any given element.

The paucity of evolutionarily informative data also reflects methodological challenges that make it difficult to distinguish functional from nonfunctional changes in the same *cis*-regulatory element from different species. Because *cis*-regulatory DNA can better tolerate changes in sequence length and the number and spacing of functional motifs, two homologous elements with a high degree of sequence divergence may nevertheless drive the same pattern of gene expression during development. A considerable body of comparative data exists for the sequences of homologous *cis*-regulatory elements obtained from multiple taxa. Although these comparisons make it possible to identify *conserved* sequences, which are often sites for critical DNA-binding proteins, it remains difficult to determine which, if any, of the remaining sequences that are not shared between elements from different species has functional and therefore potential evolutionary significance.

Despite these obstacles, some comparative studies of *cis*-regulatory elements have illustrated some general principles concerning the conservation, dynamic turnover, and functional modification of *cis*-regulatory DNA sequences in evolution. These principles are discussed next.

Conservation of functional elements among widely divergent taxa.
Given a typical mutation rate of 1 nucleotide change per 10^6 bp per generation, it can be calculated that for any animal taxa, sequences that are not under any functional constraint (that is, selection) will diverge completely in a span of several million generations. Thus comparisons of sequences from sufficiently distant taxa can reveal sites that are under functional constraint. It is now common practice to isolate homologs of genes of interest from multiple taxa. Comparisons of their noncoding regions often prove helpful in identifying probable functional *cis*-regulatory sequences.

For example, comparisons of the mouse, chicken, and pufferfish *Hoxb1* gene regulatory regions have revealed sequences that are shared among these three classes of vertebrates. In elements that control gene expression in the r4 rhombomere, four common sequence blocks include three binding sites for the Hoxb1/Pbx protein complex (Fig. 7.2). These shared sites indicate that *Hoxb1* autoregulation has been controlled by this element throughout vertebrate evolution.

Similar comparisons have been carried out for the *cis*-regulatory elements of a wide variety of toolkit genes. Although these investigations often identify a pattern of small islands of sequence conservation in elements from long-diverged taxa, some elements demonstrate a remarkable degree of sequence conservation. For example, the element that controls

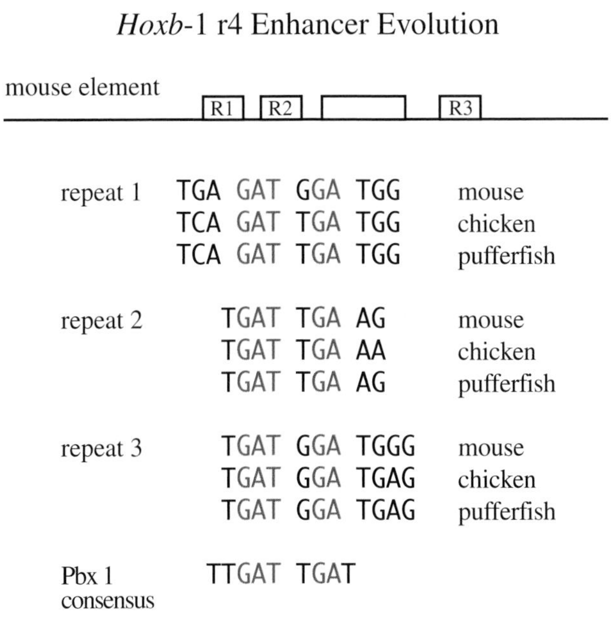

Hoxb-1 r4 Enhancer Evolution

mouse element

| R1 | R2 | | R3 |

repeat 1	TGA GAT GGA TGG	mouse
	TCA GAT TGA TGG	chicken
	TCA GAT TGA TGG	pufferfish

repeat 2	TGAT TGA AG	mouse
	TGAT TGA AA	chicken
	TGAT TGA AG	pufferfish

repeat 3	TGAT GGA TGGG	mouse
	TGAT GGA TGAG	chicken
	TGAT GGA TGAG	pufferfish

| Pbx 1 consensus | TTGAT TGAT | |

Figure 7.2
Conservation of a *Hox* gene *cis*-regulatory element in vertebrates

The *Hoxb1* r4 element controls gene expression in rhombomere 4 of several vertebrates. Within this element in the mouse, chicken, and puffer-fish, there are three highly conserved repeated sequences (R1–R3) that contain binding sites for the Pbx co-factor. The presence of these repeated sequences indicates that strong selective pressures have conserved them throughout a long period of vertebrate evolution.

Source: Data from Pöpperl H, Blenz M, Studer M, et al. Cell 1995;81:1031–1042.

expression of the *Drosophila melanogaster vestigial* gene along the dorsoventral boundary of the imaginal wing disc (see Chapter 3, Fig. 3.13) exhibits extensive sequence conservation with its counterpart in *Drosophila virilis*. In one stretch, 119 bases are perfectly conserved (Fig. 7.3). Given the approximately 60 million years during which these species have been diverging, this finding reveals great functional constraint affecting a long contiguous span of *cis*-regulatory DNA.

Some *cis*-regulatory elements may even exhibit conservation between phyla. For example, the *labial* gene of *Drosophila*, a member of the same *Hox* paralogy group as *Hoxb1*, is autoregulated by a Lab/Exd protein complex. The similar functional sites in the autoregulatory elements of vertebrate *Hoxb1* genes and the *Drosophila lab* gene could be due to conservation of an autoregulatory element from their last common ancestor.

The dynamics of sequence turnover in* cis-*regulatory DNA. The vast evolutionary distances between the taxa in the preceding examples allow us to confirm the significance of the conservation of particular sequences; they do not, however, elucidate the dynamics of *cis*-regulatory DNA evolution. For a better picture of that process, we must compare a greater number of more closely related species.

As an example, consider the homologs of the well-studied *Drosophila melanogaster even-skipped* stripe 2 element (Fig. 7.4a), which have been isolated from a host of other *Drosophila* species. These elements all drive an accurately positioned stripe of reporter gene expression in the *eve* stripe 2 domain in *D. melanogaster*, yet considerable sequence divergence occurs among the elements. Interestingly, some of the sequence changes in the other *Drosophila* species abolish sites that are known to be essential in the *D. melanogaster* element (Fig. 7.4b). For instance, the third Bicoid binding site (bcd-3) and the

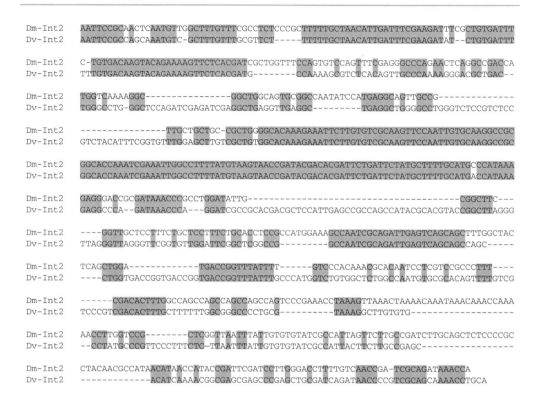

```
Dm-Int2   AATTCCGCAACTCAATGTTGGCTTTGTTTCGCCTCTCCCGCTTTTTGCTAACATTGATTTCGAAGATTTCGCTGTGATTT
Dv-Int2   AATTCCGCCAGCAAATGTC-GCTTTGTTTGCGTTCT-----TTTTTGCTAACATTGATTTCGAAGATAT--CTGTGATTT

Dm-Int2   C-TGTGACAAGTACAGAAAAGTTCTCACGATCGCTGGTTTCCAGTGTCCAGTTTCGAGGGCCCAGAACTCAGGCCGACCA
Dv-Int2   TTTGTGACAAGTACAGAAAAGTTCTCACGATG--------CCAAAAGCGTCTCACAGTTGCCCAAAAGGGACGCTGAC--

Dm-Int2   TGGTCAAAAGGC---------------GGCTGGCAGTGCGGCCAATATCCATGAGGCAGTTGCCG--------------
Dv-Int2   TGGGCCTG-GGCTCCAGATCGAGATCGAGGCTGAGGTTGAGGC---------TGAGGCTGGGGCCTGGGTCTCCGTCTCC

Dm-Int2   ---------------TTGCTGCTGC-CGCTGGGGCACAAAGAAATTCTTGTGTCGCAAGTTCCAATTGTGCAAGGCCGC
Dv-Int2   GTCTACATTTCGGTGTTTGGAGCTTGTCGCTGTGGCACAAAGAAATTCTTGTGTCGCAAGTTCCAATTGTGCAAGGCCGC

Dm-Int2   GGCACCAAATCGAAATTGGCCTTTTATGTAAGTAACCGATACGACACGATTCTGATTCTATGCTTTTGCATGCCCATAAA
Dv-Int2   GGCACCAAATCGAAATTGGCCTTTTATGTAAGTAACCGATACGACACGATTCTGATTCTATGCTTTTGCATGACCATAAA

Dm-Int2   GAGGGACCGCGATAAACCCGCCTGGATATTG--------------------------------------CGGCTTC---
Dv-Int2   GAGGCCCA--GATAAACCCA---GGATCGCCGCACGACGCTCCATTGAGCCGCCAGCCATACGCACGTACCGGCTTAGGG

Dm-Int2   ----GGTTGCTCCTTTCTGCTCCTTTCTGCACCTCCGCCATGGAAAGCCAATCGCAGATTGAGTCAGCAGCTTTGGCTAC
Dv-Int2   TTAGGGGTTAGGGTTCGGTGTTGGATTCGGCTCGGCCG---------GCCAATCGCAGATTGAGTCAGCAGCCAGC-----

Dm-Int2   TCAGCTGGA-------------TGACCGGTTTATTTT------GTCCCACAAACGCACAATCCTCGTCCGCCCTTT----
Dv-Int2   ----CTGGTGACCGGTGACCGGTGACCGGTTTATTTGCCCATGGTCTGTGGCTCTGGCCAATGTGCGCACAGTTTTGTCG

Dm-Int2   ------CGACACTTTGGCCAGCCAGCCAGCCAGCCAGTCCCGAAACCTAAAGTTAAACTAAAACAAATAAACAAACCAAA
Dv-Int2   TCCCGTCGACACTTTGCTTTTTTGGCGGGCCCCTGCG---------TAAAGGCTTGTGTG------------------

Dm-Int2   AACCTTGGTCCG--------CTCGGTTAATTTATTGTGTGTATCGCCATTAGTTCTTGCCGATCTTGCAGCTCTCCCCGC
Dv-Int2   --CCTATGCCCGTTCCCTTTCTC-TTAATTTATTGTGTGTATCGCCATTACTTCTTGCCGAGC----------------

Dm-Int2   CTACAACGCCATAACATAACCATACCGATTCGATCCTTGGGACCTTTTGTCAACCGA-TCGCAGATAAACCA
Dv-Int2   -------------ACATCAAAACGGCGAGCGAGCCCGAGCTGCGATCAGATAACCCGTCGCAGCAAAACCTGCA
```

Figure 7.3

Extensive conservation of a _cis_-regulatory element between highly diverged _Drosophila_ species

The boundary enhancer from the _vestigial_ gene of _D. melanogaster_ and _D. virilis_ controls gene expression on the dorsoventral boundary of the developing wing field. Alignment of the sequences of the two elements reveals extensive regions of perfect sequence conservation. This pattern of "islands" of sequence conservation between _cis_-regulatory elements is typical of functionally conserved elements between well-diverged species, although the length of certain conserved regions is extraordinary in this example.

Source: Williams JA, Paddock SW, Vorwerk K, and Carroll SB. Nature 1994;368:299–305.

first Hunchback (hb-1) binding site found in _D. melanogaster_ lack counterparts in _D. pseudobscura_, _D. erecta_, and _D. yakuba_. The bcd-3 site is required for proper _eve_ stripe 2 formation in _D. melanogaster_. This analysis suggests that the bcd-3 site may be a new site in _D. melanogaster_, whose function became essential in the context of other changes within the element.

Of the 17 known binding sites for transcriptional regulators in this element in _D. melanogaster_, only 3 are perfectly conserved among the other _Drosophila_ species studied (see Fig. 7.4b). Thus these binding sites exhibit a considerable degree of substitution. As each full element from the heterologous species drives an accurate _eve_ 2 stripe in _D. melanogaster_, it appears that the various changes in binding sites compensate for one another, such that the net output of the element is the same (in _D. melanogaster_).

One evolutionary explanation for the observed pattern of functional conservation and sequence changes in the *eve* stripe 2 enhancer is that **stabilizing selection** governs the dynamics of sequence turnover in the element. Because multiple sites exist for each regulatory protein, as well as both positive and negative inputs, slightly deleterious individual

a

b

Conservation of Transcription Factor Binding Sites in Eve Stripe 2 Enhancer

Sites

Species	Kr						Bcd					Hb			Gt		
	1	2	3	4	5	6	1	2	3	4	5	1	2	3	1	2	3
D. simulans	P	P	P	P	P	P	P	P	S	S	P	S	P	P	P	P	S
D. yakuba	P	P	S	P	P	P	S	P	W	S	P	S	S	P	P	S	S
D. erecta	P	P	S	P	P	P	S	P	S	S	P	S	S	P	P	S	S
D. pseudobscura	S	S	S	S	P	P	S	S	A	S	P	A	P	S	S	W	S
D. picticornis	W	S	W	S	P	P	W	S	A	S	P	A	S	W	W	W	W

Conservation

P = perfect
S = strong (1–2 changes)
W = weak (3 or more changes)
A = absent

Figure 7.4

Evolutionary dynamics of transcription factor-binding site evolution in a conserved *cis*-regulatory element

(**a**) Binding sites for the Krüppel (Kr), Giant (Gt), Bicoid (Bcd), and Hunchback (Hb) proteins in the 670 bp *D. melanogaster eve* stripe 2 *cis*-regulatory element are shown. (**b**) The conserved binding sites in five different *Drosophila* species are tabulated. The degree of sequence conservation within each site is indicated (P, S, W, A). Note that certain sites such as Kr5, Kr6, and Bcd5 are perfectly conserved, whereas other sites such as Hb1 and Bcd3 are absent from certain species.

Sources: Data from Stanojevic D, Small S, and Levine M, et al. Science 1991;254:1385–1387; Ludwig MZ, Patel NH, Kreitman M. Development 1998; 125:949–958.

changes may accumulate if compensatory mutations are also occurring. The substitution patterns seen in the *eve* elements are similar to those observed in other complex elements, suggesting that these inferences regarding the balancing effects of mutation and selection are likely to apply, in general, to the functional conservation of *cis*-regulatory elements.

*Functional modification of **cis**-regulatory DNA sequences.*   Many of the comparative studies presented in Chapters 5 and 6 inferred a change in *cis*-regulatory functions but, with the exception of *Hoxc8* element evolution in vertebrates (see Fig. 5.7), the specific functional changes within elements have not been identified. Nevertheless, some well-analyzed cases of the evolution of tissue-specific *cis*-regulatory elements do illustrate the likely mechanisms underlying the functional modification of such elements of developmental regulatory genes.

Co-option or modification of *cis*-regulatory elements is the most likely explanation for the evolution of lens crystallins (a diverse group of water-soluble proteins that have been recruited to function in eye lenses in the refraction of light onto the retina). Various enzymes or proteins related to enzymes such as lactate dehydrogenase B, argininosuccinate lyase, α-enolase, glutathione-S-transferase, and small heat-shock proteins are expressed at very high levels in the lens in particular species. The recruitment of these proteins as crystallins has occurred independently in various lineages, suggesting that many gene co-option events have occurred during the evolution of lens crystallins.

Analyses of *cis*-regulatory elements of a wide variety of crystallin genes have revealed that certain transcription factors are frequently involved in high levels of protein expression in the lens. They include the Pax-6 and retinoic acid receptor proteins, which play major roles in eye development in most animal phyla. The implication of these proteins in crystallin gene regulation suggests a scenario for the molecular genetic basis of lens crystallin recruitment. Namely, the evolution of sites for Pax-6, retinoic acid receptors, and other transcription factors within extant *cis*-regulatory elements of these genes may lead to abundant levels of expression in developing eye tissue. In general, high levels of gene expression in a given cell type or tissue could occur through the evolution of sites for abundant transcription factors within extant *cis*-elements.

Tissue-specific losses of gene expression also occur. One well-analyzed case involves the glucose dehydrogenase (*gld*) enzyme, which is expressed in particular reproductive tissues of many *Drosophila* species. In *D. teissieiri*, for example, *gld* expression does not occur in a subset of these tissues. Comparisons of the homologous *cis*-regulatory regions of the *gld* gene of seven different *Drosophila* species have revealed extensive sequence conservation between all seven genes, including the perfect conservation of certain motifs between all seven species. On the other hand, the *D. teissieiri gld* element lacks any copies of a sequence motif that occurs three times in a *D. melanogaster gld* regulatory element. This motif directs gene expression in *D. melanogaster* in the tissues from which *gld* expression in *D. teissieiri* is absent. Thus evolutionary changes in the *D. teissieiri gld* element led to tissue-specific loss of *gld* expression in this lineage.

THE ROLE OF *CIS*-REGULATORY DNA IN MORPHOLOGICAL VARIATION AND THE RESPONSE TO SELECTION

All of the examples in Chapters 5 and 6, and most of the molecular comparisons of *cis*-regulatory elements described in this chapter, have focused on differences between groups at various—often higher—taxonomic levels. Because the source of morphological evolution is morphological *variation*, however, it is important to understand the genetic regulatory architecture and the molecular basis of morphological variation *within* species. The most successful approaches to understanding intraspecific variation exploit genetic methodologies for assessing the number and identity of genes involved in morphological differences between populations. This entirely different experimental approach to morphological variation also points to a central role for *cis*-regulatory DNA in morphological evolution.

One of the better-analyzed examples of intraspecific variation comprises the bristle pattern of adult fruit flies. *Drosophila melanogaster* is covered with sensory bristles that serve mechano- and chemoreception functions. The number and pattern of bristles on various body parts can vary between individuals as well as between populations. Much is known about the genes that control the bristle pattern and their functions in the developmental processes that give rise to it. Consequently, this system offers great potential for exploiting our emerging knowledge of developmental genetics so as to better understand morphological variation.

Bristles are examples of **quantitative traits** that vary continuously in populations. Other quantitative traits include features such as body size, organ size, and life span. Quantitative genetic techniques seek to elucidate the number and identity of those loci [quantitative trait loci (QTLs)] that contribute the bulk of the genetic variance underlying the morphological variance in characters under study. Discrimination is made between genes of "small effect," many of which may combine to account for some portion of the variance in a trait, and genes of "large effect," which may account for 5% to 10% (or more) of the variance in a given trait.

Studies of *Drosophila* bristle variation have shown that many loci affect bristle number but only a few loci have large effects that cause most of the variation. Some QTLs are genes known to be involved in bristle patterning, including the *achaete-scute, scabrous,* and *Delta* genes. The *achaete* and *scute* genes encode transcription factors required for neural precursor formation, whereas the *Delta* and *scabrous* genes encode ligands involved in the *Notch*-mediated signaling pathway that regulates cell interactions in proneural clusters. Because mutations in these genes have dramatic effects on bristle development, their identification as QTLs in bristle number variation implicates them as playing a role in the evolution of bristle number.

The variation at all three loci appears to occur within regulatory regions. In the *achaete-scute* region, for instance, DNA insertions are strongly associated with variation in bristle number in natural populations. These insertions do not disrupt the protein coding regions, but they do appear to affect the expression of the *achaete* and *scute* genes. Two sites within the *Delta* gene are also associated with bristle variation. Each of these sites is located in an intron, rather than a coding segment of the gene. Similarly, sites associated with bristle variation in the *scabrous* gene appear within particular regulatory regions. These studies of *Drosophila* bristle variation have also revealed that the molecular basis for the genetic variation in loci that contribute to morphological variation may involve two or more sequence differences in a gene—not just a single difference.

Another experimental approach to studying the genetics of morphological variation is through **artificial selection** on traits of interest. Rather than relying solely on the expressed quantitative variation present in natural populations, it is possible to derive populations with much greater divergence in traits through repeated selection over several generations for individuals with character states at either end of a continuum. For example, repeated selection and breeding of individual butterflies with larger or smaller wing eyespots has established lines of butterflies with greatly divergent eyespot patterns (Fig. 7.5).

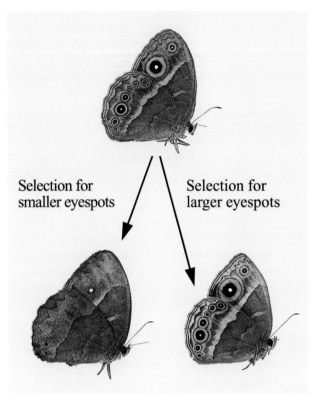

Figure 7.5
Artificial selection for morphological traits

Repeated selection and breeding of individual butterflies with larger or smaller eyespots from an initial population (**top**) leads to the establishment of lines of animals with vastly smaller (**left**) or larger (**right**) eyespots.

Source: Photos courtesy of Paul Brakefield.

Selection for smaller eyespots

Selection for larger eyespots

Similar experiments have been performed to select lines of flies with greater or fewer bristles. Genetic crosses between these fly lines with "high" and "low" bristle numbers can be used to estimate the number of loci involved in the divergence and to map the quantitative contributions of genetic regions and individual loci. Many of the candidate loci identified in these experiments encode proteins known to play roles in the patterning and genesis of bristles. Although the precise molecular localization of the genetic differences responsible for the morphological divergence remains unknown, we do understand one crucial finding in both natural and artificially selected populations. That is, many genetic loci typically contribute to differences in such modest traits as eyespot size on butterfly wings or bristle number on a single fruit fly body part. Because the products of these genes interact through regulatory networks, the effects of genetic variation are usually not strictly additive. Combinatorial interactions—so powerful in developmental processes—clearly play major roles in morphological variation and evolution.

Cryptic genetic variation and the potential for morphological evolution

The importance of gene interactions to morphological variation has also been underscored by some fascinating studies of the phenotypes that arise when mutations are introduced into individuals with different genetic backgrounds. While a single mutation is described as causing a particular phenotypic change, this relationship, in practice, depends on the context of the genetic background in which a mutation occurs.

For example, it is well established that in laboratory lines of *Drosophila melanogaster,* introduction of dominant mutations in the Sevenless tyrosine kinase receptor and EGF receptor (DER) induces a roughening of the surface of the adult eye. This outcome occurs because of a perturbation of developmental events in eye patterning that are dependent on the Sev and DER pathways. However, introduction of these mutations into wild-type flies with different genetic backgrounds produces a considerable range of severity in phenotypes. In some genetic backgrounds, the mutant phenotypes are suppressed as compared to their effects in laboratory strains; in others, they are enhanced. The enhancement observed in some backgrounds sometimes exceeded that caused by mutations in additional components of the Sev or EGF-R signal transduction pathways. This finding indicates that considerable genetic variation occurs in loci affecting the function of major pathways.

Similar observations have been made regarding the introduction of homeotic mutations into different genetic backgrounds. Both extreme modification and suppression occurs. In some cases, this phenotypic variation reflects the presence of single genes of large effect that modify the homeotic phenotype, but have no stand-alone effect on wild-type development.

Such studies demonstrate that there is widespread genetic variation that may sometimes have effects that are not discernible unless certain other interacting mutations are present. Evolutionary and developmental geneticists are only now beginning to appreciate this "cryptic" variance. The important implication of this cryptic variation is that underlying phenotypic stability, the quantitative aspects of genetic regulatory inputs may vary extensively. Furthermore, it suggests that the standing genetic variation available for the selection of new phenotypes in evolution may be significantly greater than previously thought.

One striking illustration of this latter idea has come from artificial selection for homeotic phenotypes in *Drosophila.* It has long been known that homeotic phenotypes can sometimes be obtained by environmental insults to embryos during sensitive periods. Treatment of developing wild-type flies with ether vapor, for example, can induce mimics or **phenocopies** of homeotic transformations caused by mutations in the *Ubx* gene. C. H. Waddington showed decades ago that, if one repeatedly selected for individuals demonstrating the bithorax phenocopy, the resulting populations would show a greater frequency of response to the treatment, including individuals that displayed the phenotype independent of treatment. The selection therefore uncovered genetic variation that affects development.

Developmental and molecular genetic analysis of the *Ubx* gene response to selection for the ether-induced bithorax phenocopy revealed that ether induces loss of *Ubx* expression in patches of cells in the haltere imaginal disc. In selected individuals, the frequency of lost gene expression increases and is heritable. Therefore, genetic differences must exist, either in *cis* or *trans* to the *Ubx* gene, that influence the susceptibility of the *Ubx* gene to loss of expression following ether exposure. One site that responded to selection for phenocopy induction was mapped downstream of the *Ubx* coding region in a large *cis*-regulatory region. This finding suggests that genetic variation in a *cis*-regulatory element of the *Ubx* gene affects the fly's susceptibility to loss of gene expression in response to ether vapor.

The "cryptic" variation unmasked by this artificial selection study suggests that a reservoir of cryptic variation among developmental genes is tapped during evolution by natural selection. Changes in the environment or introduction of a new mutation into a population can quickly unmask the cryptic genetic variation as an expressed phenotypic variation. The increase in phenotypic variation provides a broader range of fitnesses upon which natural selection can act.

THE EVOLUTION OF REGULATORY DNA AND MORPHOLOGICAL DIVERSITY

In this chapter, we have examined the evolution of regulatory DNA from three perspectives —theoretical considerations of *cis*-regulatory element function, comparative analyses of known *cis*-regulatory elements, and a handful of genetic investigations into the nature of morphological variation. All of these approaches support the claim that regulatory DNA is the source of genetic diversity that underlies morphological diversity.

The greater ability of regulatory DNA to evolve (as compared to coding DNA) is enabled by three critical factors:

- The degree of freedom in *cis*-regulatory sequences
- The modularity of *cis*-regulatory elements
- The combinatorial action of the transcription factor repertoire in cells

The degree of freedom in regulatory DNA is important because it imparts a greater tolerance of regulatory DNA to all types of mutational change. Consequently, regulatory DNA does not need to maintain any reading frame, and it can function at widely varying distances from or orientation to the transcription units it controls.

The modularity of the elements that make up the *cis*-regulatory systems of genes facilitates evolution because individual elements can evolve independently. Its ability to readily evolve means that regulatory DNA function is a rich source of genetic variation and, therefore, potential morphological variation.

The importance of the combinatorial nature of transcription regulation cannot be overemphasized. The transcription factor repertoire is sufficiently diverse and the stringency of DNA binding sufficiently relaxed such that sites for most transcription factors can evolve at a significant frequency in animal genomes. As new combinations of sites arise in existing functional elements, and potentially within nonfunctional DNA as well, variations in the timing, level, and spatial domains of gene expression may evolve and generate phenotypic variations, which serve as the raw material for morphological evolution. The very structure of the *cis*-regulatory regions of toolkit genes—that is, its composition of multiple, independently regulated *cis*-elements—is the product of these evolutionary processes. It constitutes persuasive evidence that the diversification of regulatory DNA, while preserving coding function, is the most available and most frequently exploited mode of genetic diversification in animal evolution.

Extrapolating from bristles to body plans

The focus of this book has been to explore and explain the genetic basis of the morphological diversity of animals. Three kinds of assumptions are implicit to our approach and the conclusions we have drawn.

First, we assume that we can extrapolate from the present to the past. *Aysheaia* (see Fig. 1.1a) and *Acanthostega* (see Fig. 1.5e) are no longer walking the Earth, but we assume that we can extrapolate from genetic and developmental knowledge of their modern descendants and thereby infer processes that occurred in the Cambrian or the Devonian periods.

Second, we assume that we can extrapolate from the particular to the general. Very few genetic regulatory circuits or networks are known in great detail, relatively few *cis*-regulatory elements have been thoroughly dissected at the DNA sequence and transcription factor level, and even fewer regulatory elements have been compared between relevant taxa. Nevertheless, we assert that the specific knowledge we do have is likely to apply to the general case.

Our third, and perhaps most controversial, assumption is that we may extrapolate from the observable small genetic changes underlying fine-scale morphological variation to the largest changes that have happened in animal history. Jacques Monod once asserted that what was true of the bacterium *E. coli* was true of elephants. Analogously, we suggest that what is true of bristles is true of body plans. If dozens of regulatory changes and gene interactions underlie the difference of just a few bristles between fruit fly populations, then the number of genetic regulatory differences underlying the full range of morphological differences between a fly and a butterfly, or between a mouse and a human, must be staggeringly immense. But, do we need to invoke any additional or special genetic or evolutionary mechanisms beyond those illustrated for bristle variation to explain morphological diversity at higher levels?

Is there any role, for example, for the sorts of dramatic variants described by Bateson in the evolution of large differences between taxa? The emerging evidence suggests not. Although, in theory, any morphological or genetic variation may potentially be selected for, Bateson's monsters (for example, homeotic mutants) are generally less fit than other individuals and would be selected against in interbreeding populations. Furthermore, the case studies and mechanisms described in this book suggest that not only are such large steps improbable, but unnecessary. We have described many examples of morphological evolution that involved homeotic genes, but not homeotic mutations. Dramatic morphological changes, such as the fin-to-limb transition in vertebrates or the evolution of the insect haltere, involved many regulatory, developmental, and anatomical modifications that could not and did not evolve instantaneously. Instead, these structures were sculpted by regulatory evolution over millions of years.

Does variation exist in animal forms that might provide the basis for large-scale morphological evolution? Indeed, variants are found with significant frequency within and between natural populations for characters that are more significant to the evolution of body plans and body parts than just bristle number. For example, detailed analysis of a single population of newts has revealed a surprising array of variation in limb skeleton morphologies in approximately 30% of individuals examined (Fig. 7.6). Most interestingly, these variant patterns often resembled the standard limb skeletal morphologies found in other species. Some of these patterns were similar to more "ancient" patterns; others reflected more derived conditions. These observations suggest that within this single population of animals, many potential limb morphologies are expressed that represent both potential novelties and atavisms (that is, a return to an ancestral state). As the evolution of limb skeletal morphology is important to the functional adaptation of these and other tetrapods, the variation documented in this population suggests that the evolution of new forms or the reversion to "old" forms can arise through readily available genetic variation in the genes that affect limb morphology. Little, if any, morphological variation in this population is likely to be due to intraspecific variation in protein function.

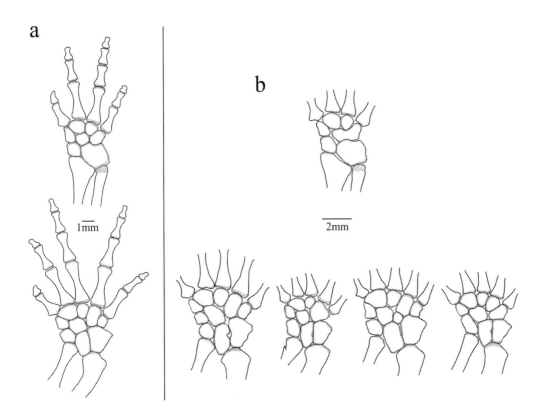

Figure 7.6
Intraspecific variation in limb morphology in salamanders

A single population of salamanders displays considerable variations in limb morphology. (**a**, **top and bottom**) The standard bone patterns of the forelimb and hindlimb, respectively, of *Taricha granulosa*. (**b**) Variations of the standard patterns. (**top**) A forelimb bone pattern in which fewer distinct carpal elements form. (**bottom**) Four variants in hindlimb patterns in which the number and pattern of carpal structures differ.

Source: Shubin N, Wake DB, Crawford AJ. Evolution 1995;49:874–884.

Other well-documented examples of striking intraspecific morphological variation have been described in centipedes and certain fishes. In geophilomorph centipede species, the number of leg-bearing segments varies from 29 to 191; in other centipede orders, this number remains relatively constant. Within geophilomorph species, there may be a range of variance of 12 to 14 segments. Given the considerable range in segment number observed in fossil and extant arthropods, the existence of mechanisms underlying variation in such a central feature of the arthropod body plan is of immense interest.

Remarkable body pattern variation is also observed between populations of the three-spine stickleback fish. This group of species has experienced repeated postglacial episodes of colonization of freshwater lakes and streams. Skull, body, and appendage shapes differ extensively between populations (Fig. 7.7) that have become isolated relatively recently

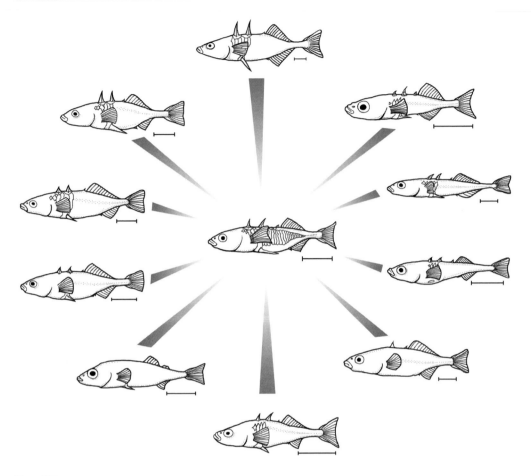

Figure 7.7
Variation in stickleback fish body patterns

General body form, fin morphology, dorsal spine number, coloration, and several other morphological features differ considerably between populations of the three-spine stickleback *Gasterosteus aculeatus* species complex. Lake and river populations from various North American populations are depicted around the periphery of a representative marine form. The striking variation of recently isolated species in this complex illustrates the potential for rapid morphological diversification of major body characters.

Source: Bell MA, Foster SA. In: Bell MA, Foster SA, eds. The evolutionary biology of the threespine stickleback. Oxford, UK: Oxford University Press, 1994:1–27.

(for example, circa 13,000 years ago). It appears that, in the stickleback, colonization and ecological diversification have led to a radiation of morphologically diverse, but closely related species.

These surveys of natural populations offer striking support for the idea that the morphological variation required for the evolution of new body patterns is available, at least in some groups. When this information is coupled with new insights into the scope of genetic variation that can lurk beneath the surface of phenotypic stability, it appears that the developmental, genetic, and potential morphological diversity of any large interbreeding group is much greater than has been realized by biologists, or is realized in the course of evolution. This potential is generated from the creative power of regulatory change and interaction; it is realized, of course, through ecological interactions across many levels of biology.

We arrive at a view—while perhaps made more sophisticated by virtue of modern embryology, genetics, and molecular biology—that is not far from the spirit, if not the heart, of Darwin's original ideas. Darwin chose to open *The Origin of Species* with a discussion of domesticated species, arguing persuasively about the power of selection upon variation, on a scale and landscape familiar to his readers. Then, in closing his great work, he urged his audience to extrapolate from dog and pigeon breeding to the larger landscape of life's history:

> "we are always slow in admitting great changes of which we do not see the steps . . . . The mind cannot possibly grasp the full meaning of the term of even a million years; it cannot add up and perceive the full effects of many slight variations, accumulated during an almost infinite number of generations."

Today, a similar challenge applies to our attempts to identify the genetic steps underlying the diversification of any group, let alone to conceive and reconstruct the innumerable steps underlying the great changes in animal designs that have unfolded from our Precambrian ancestors. Only recently have we been able to perceive some of the general mechanisms at work. Armed with increasingly more powerful tools for analyzing variation and comparing genomes and gene expression, the relationship between the evolution of DNA and animal diversity is drawing increasingly within our reach.

SELECTED READINGS

Regulatory Evolution: Early Ideas

Britten, R. J., Davidson, E. H. Repetitive and non-repetitive DNA sequences and a speculation on the origins of evolutionary novelty. *Qtly Rev Biol* 1971; 46:111–138.

Jacob, F. Evolution and tinkering. *Science* 1977; 196:1161–1166.

Wilson, A. C. The molecular basis of evolution. *Sci Am* 1985; 253:164–173.

Zuckerkandl, E., Pauling, L. Evolutionary divergence and convergence in proteins. In Bryson, V., Vogel, J. H., eds. Evolving genes and proteins. New York: Academic Press, 1965: 97–166.

Functions of cis-Regulatory Elements and Transcription Factor Interactions

Arnone, M. I., Davidson, E. H. The hardwiring of development: organization and function of genomic regulatory systems. *Development* 1997; 124:1851–1864.

Lin, Y.-S., Carey, M., Ptashne, M., Green, M. R. How different eukaryotic transcriptional activators can cooperate promiscuously. *Nature* 1990; 345:359–361.

Ptashne, M. A genetic switch: phage lambda and higher organisms, 2nd ed. Cambridge, MA: Blackwell Scientific, 1992.

Schüle, R., Muller, M., Kaltschmidt, C., Renkawitz, R. Many transcription factors interact synergistically with steroid receptors. *Science* 1988; 242:1418–1420.

Evolution of cis-Regulatory Elements

Ludwig, M. Z., Bergman, C., Patel, N. H., Kreitman, M. Evidence for stabilizing selection in a eukaryotic enhancer element. *Nature* 2000; 403:564–567.

Ludwig, M. Z., Patel, N. H., Kreitman, M. Functional analysis of eve stripe 2 enhancer evolution in *Drosophila:* rules governing conservation and change. *Development* 1998; 125:949–958.

Pöpperl, H., Blenz, M., Studer, M., et al. Segmental expression of *Hoxb-1* is controlled by a highly conserved autoregulatory loop dependent upon *exd/pbx. Cell* 1995; 81:1031–1042.

Ross, J. L., Fong, P. P., Cavener, D. R. Correlated evolution of the cis-acting regulatory elements and developmental expression of the *Drosophila Gld* gene in seven species from the subgroup *melanogaster. Develop Genetics* 1994; 15:38–50.

Intraspecific Variation: Quantitative and Developmental Genetics

Gibson, G., Wemple, M., van Helden, S. Potential variance affecting homeotic *Ultrabithorax* and *Antennapedia* phenotypes in *Drosophila melanogaster. Genetics* 1999; 151:1081–1091.

Lai, C., Lyman, R. F., Long, A. D., et al. Naturally occurring variation in bristle number and DNA polymorphisms at the *scabrous* locus of *Drosophila melanogaster. Science* 1994; 266:1697–1702.

Long, A. D., Lyman, R. F., Langley, C. H., Mackay, T. F. C. Two sites in the *Delta* gene region contribute to naturally occurring variation in bristle number in *Drosophila melanogaster. Genetics* 1998; 149:999–1017.

Mackay, T. F. C., Langley, C. H. Molecular and phenotypic variation in the *achaete-scute* region of *Drosophila melanogaster. Nature* 1990; 348:64–66.

Mackay, T. F. C. The nature of quantitative genetic variation revisited: lessons from *Drosophila* bristles. *BioEssays* 1996; 18:113–121.

Polaczyk, P. J., Gasperini, R., Gibson, G. Naturally occurring genetic variation affects *Drosophila* photoreceptor determination. *Dev Genes Evol* 1998; 207:462–470.

Artificial Selection

Brakefield, P. M., Gates, J., Keys, D., et al. The development, plasticity and evolution of butterfly eyespot patterns. *Nature* 1996; 384:236–242.

Gibson, G., Hogness, D. S. Effect of polymorphism in the *Drosophila* regulatory gene Ultrabithorax on homeotic stability. *Science* 1996; 271:200–203.

Nuzhdin, S. V., Dilda, C. L., Mackay, T. F. C. The genetic architecture of selection response: inferences from fine-scale mapping of bristle number quantitative trait loci in *Drosophila melanogaster. Genetics* 1999; 153:1317–1331.

Wang, R-L., Stec, A., Hey, J., et al. The limits of selection during maize domestication. *Nature* 1999; 398:236–239.

Morphological Variation within Species

Anh, D-G., Gibson, G. Axial variation in the threespine stickleback: genetic and environmental factors. *Evol Develop* 1999; 1:100–112.

Arthur, W. Variable segment number in centipedes: population genetics meets evolutionary developmental biology. *Evol Develop* 1999; 1:62–69.

Shubin, N., Wake, D. B., Crawford, A. J. Morphological variation in the limbs of *Taricha granulosa* (Caudata: Salamandridae): evolutionary and phylogenetic implications. *Evolution* 1995; 49:874–884.

Extrapolation from Microevolution to Macroevolution

Akam, M. *Hox* genes, homeosis and the evolution of segment identity: no need for hopeless monsters. *Intl J Develop Biol* 1998;4 2:445–451.

Leroi, A. M. The scale independence of evolution. *Evol Develop* 2000; 2:67–77.

Stern, D. L. A role of *Ultrabithorax* in morphological differences between *Drosophila* species. *Nature* 1998; 396:463–466.

Wallace, B. Reflections on the still "hopeful monster." *Qtly Rev Biol* 1985; 60:31–42.

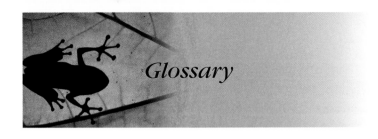

Glossary

activator A protein that positively regulates transcription of a gene.

anterior group *Hox* genes *Hox* genes expressed in the anterior region of bilaterians and located toward the 3′ end of *Hox* clusters; include the *Hox1* and *Hox2* genes in vertebrates.

anural The tailless condition in chordates.

apical ectodermal ridge (AER) A thickening of the ectoderm at the distal extent of developing tetrapod limbs.

artificial selection Selection imposed by humans for one or more traits, usually in a laboratory, captive, or domesticated population.

atavism The evolutionary reversion of a character to an ancestral state.

autopod The distal portion of the vertebrate limb, including the hand/foot and digits.

axes The polarity of animals can be defined in three dimensions. In bilaterians, the two main axes are the *anteroposterior* (head-to-tail or *rostrocaudal*) axis and the orthogonal *dorsoventral* axis. projections from the body wall, such as appendages, also possess a third *proximodistal* axis.

basal In phylogenetic terms, a group that branches at the base of a clade.

battery A group of genes regulated by the same transcription factor.

Bilateria The bilaterally symmetrical animals, including all protostomes and deuterostomes, but not sponges, cnidarians, or ctenophores.

blastopore An invagination on the surface of embryos into which the mesoderm and endoderm move during early embryogenesis.

Cambrian The geological period from 544 to 490 million years ago, during which the diversity of modern animal phyla expanded.

cell autonomous Genes whose products affect only the differentiation or behavior of cells in which they are expressed. Contrast with genes that encode signaling ligands that act on other cells (i.e., non-cell-autonomously).

central group *Hox* genes *Hox* genes expressed in the central region of bilaterians and located in the middle of *Hox* clusters; include the *Hox3* through *Hox8* genes in vertebrates.

cephalochordate Phylum composed of only approximately 25 marine species characterized by a notochord and small anterior brain vesicle. Also known as the lancelets, due to their body shape, they constitute the nearest phylogenetic outgroup to the vertebrates.

chromatin The material into which DNA in the cell nucleus is packaged with proteins.

***cis*-regulatory element** A discrete region of DNA that affects transcription of a gene.

clade A group of species descended from a common ancestral species. Also known as a monophyletic group.

Cnidaria Phylum comprising approximately 8000 living, mostly marine species, all of which possess nematocysts. Their diploblastic body organization consists of an outer cell layer (ectoderm) and an inner cell layer (endoderm). Cnidaria includes jellyfish, sea anemones, and corals.

co-activator Protein that interacts with the activator in control of transcription. Some co-activators possess activities that modify local chromatin conformation.

colinearity The correlation between the order of *Hox* genes on a chromosome and the rostrocaudal order of gene deployment in the embryo.

combinatorial regulation The control of gene transcription by two or more transcription factors. The spatial patterns of gene expression are often delimited by the combined action of transcription factors.

compartment A discrete subdivision of a developing field that contains populations of cells that share a lineage restriction. Cells of one compartment do not intermix with adjacent cells in other compartments.

compensatory mutation A mutation that restores or maintains the function of a gene or *cis*-regulatory element by counteracting the effects of one or more different mutations.

cooperativity The enhanced binding of a protein to a *cis*-regulatory DNA sequence due to interaction with other bound proteins.

co-option The recruitment of genes into new developmental or biochemical functions.

corepressor A protein that interacts with a repressor in the control of transcription. Corepressors may modify the local conformation of chromatin.

cryptic genetic variation The genetic variation in components of developmental programs that exists in the absence of phenotypic variation. Perturbations to developmental programs may unmask this variation and lead to phenotypic variation.

deuterostomes The group of animals including echinoderms, ascidians, cephalochordates, and vertebrates in which the mouth forms at a site separate from the blastopore.

Devonian The geological period from approximately 409 to 362 million years ago, during which terrestrial forms invaded the land.

diploblasts Animals possessing two cell layers—the outer ectoderm layer and the inner endoderm layer; includes cnidarians.

disparity The variety of designs in body plans.

diversity The number of species in a clade; the divergence of morphology.

Ecdysozoa A great clade of protostomes including the arthropods, nematodes, onychophora, and priapulids. Named for the shared character of molting.

echinoderm A deuterostome animal with a pentaradially symmetrical adult body plan.

ectoderm The outer tissue layer of animal embryos that gives rise to the epidermis and the nervous system.

Ediacara The fauna that lived 575 to 544 million years ago, named for their fossil deposits in the hills of South Australia.

endoderm The inner tissue layer of animal embryos that gives rise to the gut and other associated organs

exon shuffling The process through which new genes are generated by recombination of one or more exons of other genes.

extant Related to living species.

field-specific selector gene A specific class of selector genes that control the formation and patterning of morphogenetic fields such as the eye, leg, and wing.

focus The developmental organizer at the center of the butterfly eyespot field.

gap genes The genes that subdivide the *Drosophila* embryo into several regions encompassing many segments. They encode transcription factors that regulate the expression of other gap genes, pair-rule genes, and homeotic genes.

gene complex A group of adjacent genes related by gene duplication.

gene divergence The process through which two or more genes created by a duplication event(s) acquire distinct functions. Changes in regulatory and/or coding regions lead to differences in gene function.

gene duplication The creation of additional genes from the template of one gene.

gene family Two or more genes with related sequences that are derived from a common ancestral gene. They are not necessarily linked, may be widely dispersed in the genome, and may be greatly diverged in function.

genetic redundancy Two genes are redundant when the absence of the function of one gene causes little or no effect because a second gene can operate in its place.

germ layers The regions of early animal embryos that give rise to the different tissue layers. Bilaterians have three germ layers, which eventually become the ectoderm, mesoderm, and endoderm.

gradient Formed when the concentration of a regulatory substance varies with the distance from a source.

haltere The hindwing of two-winged insects such as flies and mosquitoes.

helix-loop-helix A structural motif of a distinct class of transcription factors in which the helical domains of two interacting subunits are interrupted by nonhelical polypeptide loops.

hierarchy The organization of regulatory circuits into two or more tiers. Gene products in one tier control the expression of genes in lower tiers.

homeobox A 180 bp region of DNA encoding a particular class of DNA-binding domains. Approximately 20 families of homeobox-containing genes exist.

homeodomain The protein domain encoded by the homeobox.

homeotic genes Genes that regulate the identity of body regions. Mutations in homeotic genes cause the transformation of one body region or part into the likeness of another.

homolog When referring to genes, indicates genes that share similarities in their sequence because they evolved from a common ancestral gene. When referring to morphological structures, indicates features that are shared due to common ancestry.

homonomous A body plan or body part composed of repeating, similar parts.

housekeeping genes Genes that encode proteins required for basic functions required in all cells.

***Hox* genes** Homeotic, homeobox-containing genes found in linked clusters in all bilaterians.

imaginal disc Small sets of epithelial cells that are set aside during *Drosophila* embryogenesis, subsequently proliferate in the larva, and give rise to adult structures.

individualization/individuation The process of morphological differentiation of similar body parts.

lateral plate mesoderm Mesoderm in vertebrate embryos that gives rise to the blood, kidneys, and heart tissues.

leucine zipper A common transcription factor motif in which two subunits of the protein interact through leucine-containing repeat sequences.

Lophotrochozoa A great clade of protostomes including the annelids, molluscs, flatworms, and other phyla that produce a particular form of larva (trochophore) or feeding structure (lophophore).

maternal effect Genes that are expressed during oogenesis and whose products function in the development of the embryo.

meristic variation Differences in the number of repeating structures.

mesoderm The tissue layer between the ectoderm and the endoderm that gives rise to the musculature of all bilaterians.

Metazoa Multicellular animals, including diploblasts and triploblasts.

modularity The organization of animals into developmentally and anatomically distinct parts. Also, the organization of gene regulatory regions into discrete *cis*-regulatory elements.

molecular clock Differences in the sequences of RNA, DNA, and protein molecules accumulate as species diverge. In theory, geological information can be used to calibrate the rate of sequence changes between lineages. The relative sequence divergence between species can be used to infer phylogenetic relationships.

morphogen A substance whose concentration varies across a tissue or field and to which cells respond differently at different concentrations.

morphogenetic field A discrete region of an embryo that will give rise to a structure and within which pattern formation is largely independent of other developing structures.

network Two or more regulatory circuits that are linked by regulatory interactions between components of each circuit.

neural crest A population of cells in vertebrates that migrates from the edge of the neural plate to different regions of the body and gives rise to tissues such as the facial skeleton, melanocytes, and peripheral neurons.

notochord A structure that runs from head to tail in chordates and lies beneath the developing central nervous system.

novelty New morphological characters of adaptive value.

onychophora A phylum of soft-bodied animals possessing unjointed appendages; closely related to the arthropods.

oral–aboral axis The axis orthogonal to the pentaradial arms of echinoderms.

Ordovician The geological period from approximately 490 to 439 million years ago, which followed the Cambrian and preceded the Devonian.

organizer A signaling center in a developing embryo or field that induces the development of surrounding tissues. Examples include the Spemann organizer of amphibian embryos and the focus in butterfly imaginal wing discs.

orthologs Homologous genes in different species that arose from a single gene in the last common ancestor of these species.

outgroup A taxon that diverged from a group of other taxa before those taxa subsequently diverged from one another.

pair-rule genes *Drosophila* genes that are usually expressed in seven transverse stripes across the future segmented region of the early embryo, one stripe for every two segment primordia.

***ParaHox* genes** A small cluster of genes that arose from an early duplication of a primitive *Proto-Hox* gene cluster. *ParaHox* genes function in endoderm development and may be involved in the evolution of the through gut in bilaterians.

paralogs Homologous genes that are related by duplication of an ancestral gene.

parasegment Units of the *Drosophila* embryo composed of the posterior part of one segment and the anterior part of the adjacent segment.

paraxial mesoderm That part of the vertebrate mesoderm that gives rise to the somites.

pathway In cell–cell signaling, the components required for the sending, receiving, and transduction of a signal, including one or more ligands, membrane-associated receptors, intracellular signal transducers, and transcription factors.

phenocopy A mimic of a genetic mutant phenotype, usually caused by environmental changes.

phylogenetic tree A depiction of the evolutionary relationships among species and their order of "branching" from common ancestors.

pleiotropic Genes or proteins with multiple functions.

polychaete One class of annelids distinguished by the presence of chaetae, projections from the sides of each body segment.

posterior group *Hox* genes *Hox* genes that are expressed in the posterior or caudal region of bilaterians and are found at the 5′ end of *Hox* clusters; include the *Abd-B*-related and *Hox9-13* genes.

primordia A discrete field of cells that will give rise to a particular organ, appendage, or tissue type.

proleg The leg-like abdominal appendages found on the larva of certain insects, particularly Lepidoptera.

promoter The region of a gene near the start site of transcription to which the general transcriptional machinery binds.

proneural cluster A group of cells from which one or more neural precursor cells will segregate, divide, and differentiate.

Proterozoic The geological era from approximately 2500 to 544 million years ago.

protostomes One of two clades of bilaterians characterized by the origin of the mouth at the blastopore.

pseudogene The remnant of a gene that has been rendered nonfunctional through the accumulation of mutations.

quantitative trait A character that exhibits continuous variation in a population.

quantitative trait loci (QTL) Genetic loci that contribute to the variation in a quantitative trait in a population.

radiation The evolutionary divergence of a lineage into a variety of forms; often used to describe rapid diversification.

region-specific selector One class of selector genes that regulates the identity of contiguous body regions (e.g., *Hox* genes).

regulatory circuit A signaling pathway and one or more of the target genes regulated by the pathway in a given cell, tissue, or field.

regulatory evolution Evolutionary changes in gene regulation.

reporter gene A gene whose expression is used to visualize the activity of a heterologous, linked *cis*-regulatory element in vivo.

repressor A transcription factor that negatively regulates the expression of a gene, often by binding directly to DNA sequences in a *cis*-regulatory element.

rhombomere A subdivision of the vertebrate hindbrain.

secondary field Discrete units of development that are specified in the developing embryo after the primary axes are established and that give rise to appendages and organs. Pattern formation within secondary fields occurs independently of other fields.

segment polarity gene Genes that act within developing *Drosophila* segments to regulate the anteroposterior polarity of each segment.

selector gene A gene that controls cell fate.

serial homologs Repeated structures of a single organism that share a similar developmental origin.

somite A segmented subdivision of the vertebrate mesoderm that gives rise to vertebrae and associated processes, selected muscles, and the dermis.

stabilizing selection Selection that acts to keep a character constant in a population.

syncytium An embryo or tissue containing nuclei that are not separated by cell membranes.

synteny The occurrence of two genes on the same chromosome.

tetrapod The four-limbed terrestrial classes of vertebrates, including amphibians, reptiles, birds, and mammals.

trace fossil The impressions left in sediments by the meanderings and burrowing activities of ancient animals.

transcription factor A protein that regulates the transcription of genes, often, but not exclusively, by binding to *cis*-regulatory elements.

triploblasts Animals composed of three germ layers; include all bilaterians.

Urbilateria The hypothetical last common ancestor of all bilaterians.

urochordate A marine phylum in which the notochord (or urochord) is found only in the larval tail.

urodele The tailed form of chordates.

zinc finger A distinct class of transcription factors in which a DNA-binding polypeptide loop or "finger" forms through a coordination complex of zinc with cysteine and histidine residues at the base of the loop.

zone of polarizing activity (ZPA) The organizer at the posterior margin of the developing vertebrate limb that regulates anteroposterior polarity.

zygote The fertilized egg. It contains the genetic contributions of both the male and female parents.

Index

A

abdominal-A (abd-A) gene, 22
  anterior boundaries of, 129
  in evolution of insect wings, 137–139
Abdominal-B (Abd-B) gene, 22
  in evolution of insect wings, 137–139
Abdomen, limbless, 57
  evolution of in insects, 135–136
Acanthostega, fossil tetrapod limb of, 9*f*
Achaete-Scute Complex (AS-C), 31–32, 154
Achaete-scute genes, 45–46
Activators, 52
  in gene expression patterns, 64
AER. *See* Apical ectodermal ridge (AER)
Amphibians, organizers in, 16
Ancestral expression domains, partitioning of, 105
Animal body patterns
  developmental features pushing origin of, 118–119
  diversity of, 123–124
Animal design
  embryonic organizers of, 16–18
  features of, 8–12
  genetic toolkit in, 47
Animal development
  genetic regulation of, 51–92
  protein domains crucial to, 106–107
Animal evolution
  Ediacarans in, 3
  first, toolkit assembly in, 105–108
  history of, 1–13
  milestones in history of, 1

origins of, 2–5
phylogeny of, mapping gene duplication events onto, 104–105
toolkit gene in, 112–117
Animal lineages
  divergence of, 103–104
  radiation of, 98–100
Animal phyla
  living, 6
  radiation of, 1
Animal tree, 6–8
Annelids
  body plans of, 9*f*
  in early fossil record, 4*f*
  Hox expression patterns in, 131
Antennapedia (ANT-C) complex, 22, 26
Antennapedia gene, 22
  mutations of, 20, 21*f*
Anteroposterior axis
  developmental genes for, 33–35
  diversity of in arthropods and vertebrates, 124–135
  embryonic development of, 59–62
  evolution of, 112
  limb bud outgrowth in regulatory hierarchy of, 84–87
  regulatory genes of, integrating with dorsoventral coordinate system, 68–71
  in vertebrate body plan, 79–80
Anteroposterior coordinate system, 72
  development of, 58–62
Apical ectodermal ridge (AER), 16, 17*f*, 46–47, 84, 166
  formation of, 86–87
  and *Hox* gene expression, 87
Appendages. *See also* limbs; wings; legs
  epithelial, 150–152
  paired, jointed, 57

apterous (ap) genes, 30–31
  in dorsal leg evolution, 152–154
Apterous protein, 72
Artemia, anterior/posterior axis of, 127
Arthropod/onychophoran clade, 124
Arthropods
  body plans of, 9*f*
  clades of, 8
  diversity of anterior/posterior body organization in, 124–135
  Dll and *Ubx* expression in, 136*f*
  in early fossil record, 4*f*
  evolution of segmentation in, 125–127
  fossil record of, 125
  homeotic transformations in, 19, 20*f*
  Hox gene expression patterns in head of, 130*f*
  Hox gene in formation of, 26
  meristic differences among, 11*f*
  radiation of body plans of, 98–100, 123
  shared body plan of, 124
  trunk *Hox* gene expression and body changes in, 127–129
  Ubx gene expression in, 128*f*
Artificial selection, 187–188
AS-C (MASH) genes, 44–46
Ascidian species, tailed and tailless forms of, 165*f*
Ascidian tail, loss of, 163–165
ASH1 gene
  butterfly, 155*f*
  in butterfly wing color scales, 154
Atavism, 140–141
Autopod, 160
Axial identity, 132*f*
Axial patterns, vertebrate, 131–133
Aysheaia pedunculata, 2*f*

B

Bates, Henry Walter, 149

Bateson, William, 8, 15
  homeotic specimens of, 19–20

Bats, forelimb morphology in, 143–144

Battery, 57

Bcd proteins, 60–61

Beetles
  Dll and *Ubx* coexpression in pleuropod appendages of, 137*f*
  wing morphology of, 139*f*

Bicoid gene, 111
  mutant of, 33*f*

Bicoid proteins, 35, 60, 61*f*
  binding sites, 183–184

Bicyclus hindwings, 156*f*

Bilateral symmetry, evolution of, 1, 166–167

Bilaterian body plans
  complexity of, 117
  radiation of, 119, 124

Bilaterian clades, 7*f*
  phylogeny of, 98–100

Bilaterian phyla, 3
  clades of, 6
  defining features of, 166
  Dll expression in, 118*f*
  Hox gene complex expansion in, 108–110
  primitive, 3–4
  rebuilding from phylogenetic inference, 114–117

Bird forelimb
  conserved selector genes regulating, 145*f*
  morphology of, 143–144

Bithorax, 19–20
  complex, 22
  Hox genes of, 26

Blastopore, 6

BMP signaling, 151

Bobcat gene, 164–165

Body axes, formation of, 32–43

Body parts
  diversity of, 8–12

homologous, 144–145
  modularity of, 9*f*

Body patterns
  variations of, 192–193

Body plans
  diversification of, 8, 123
  echinoderm, 166–167
  insect, 57–77
  modular organization of, 1, 9*f*
  morphological diversity within conserved, 135–144
    regulatory DNA in, 190–194
  radical changes of, 163–167
  segmentation of, 125–127

Bony fishes, fins of, 159

Brachiopoda, 4*f*

Brachyury (T) gene, 163
  recruitment of during notochord evolution, 164*f*

Brain, vertebrate
  evolution of, 161–163
  subdivision of, 162–163

Bristles
  butterfly scales evolved from, 154, 155*f*
  evolution of, 187, 190–194

Burgessochaeta setigera, 2*f*

Butterfly
  eyespots of
    color rings of, 157–158
    evolution of, 155–159
    recruiting Hh signaling pathway in, 157
  larvae of, *Ubx* expression with abdominal limbs in, 142*f*
  scales of, evolved from insect bristles, 155*f*
  wings of
    color scales, 154, 155*f*
    morphology of, 139*f*
    patterning structures of, 168

C

Cad proteins, 60–61

Cambrian period
  bilaterally symmetric animals in, 1

diversification of body plans in, 123
  fossil record of, 2–4

Carboniferous period, 152

Caudal gene, 33

Cell-type-specific selector genes, 31–32, 44–46

Cellular fields, gene expression in, 90–92

Centipedes, body segments, 11*f*

Central nervous system
  chordate, 162
  complex organization of, 163

Cephalochordates, divergence of, 110

Chelicerates, 124
  Hox gene expression patterns in head of, 130*f*

Chimpanzee, vertebrae of, 11*f*

Choanoflagellates, 7*f*

Chordates
  body plans of, 9*f*
  in early fossil record, 4*f*
  lacking chordate features, 163–165
  radiation of body plans of, 123

Chromatin, 51
  local state of, 52
  repression of, 108

Chromosomal deletions, 105

Cimbex axillaris, homeotic transformations in, 20*f*

Circuit, 56

cis-regulatory DNA
  dynamics of sequence turnover in, 183–186
  functional modification of sequences of, 186
  molecular evolution of, 177*f*
  in morphological variation and response to selection, 186–190

cis-regulatory elements, 52–53
  analysis of, 54*f*
  co-option of, 177*f*, 181
  combinatorial control of, 63*f*
  conservation of between divergent species, 184*f*

de novo evolution of, 176–180, 177*f*

duplications and DNA rearrangements of, 108, 177*f*

evolution of, 175–186

   from existing elements, 180

as evolutionary change units, 173

function of, 175

gene repression versus activation in, 181–182

in *Hoxc8* expression evolution, 133–135

integrating signaling and selector protein inputs by, 76*f*

modification of existing, 177*f*, 180–181

modularity of, 91

in partitioning of ancestral expression domains, 105

regulating *Hox* genes, 67–68

in segmentation, 126–127

in tetrapod limb evolution, 160

of *Ubx* gene, 68*f*

wing-specific, 74–75

Cnidarian, 3, 7*f*, 106

in early fossil record, 4*f*

genomes of, 107–108

Hox genes of, 108

Co-activator proteins, 52

recruitment of, 178

Co-option, *cis*-regulatory DNA, 186

Coelom, 112

Coleoptera, 141*f*

Colinearity, 22

Combinatorial control mechanisms, 64

Combinatorial regulation

in spatial patterns, 91–92

of wing patterning by signaling proteins, 74

Compartments, 18

selector genes of, 28–31

Concentration-dependent response, 64

Conservation

of body plan, morphological diversity within, 135–144

of developmental functions, 114–117

of functional elements among widely divergent taxa, 182–183

of insect and crustacean head segmentation, 130–131

of segmental *engrailed* expression, 126*f*

of transcription factor binding sites, 185*f*

Conserved genes, 98–100, 113–114

homologous structure divergence and, 145*f*

Conserved regulatory circuits, 158

Coordinate systems

generation of, 71–72

sequential generation of, 90–91

Corepressors, 52

Crustaceans, 124

body organizations of, 127–128

epipodites, similarities to insect wings in, 152

Hox gene expression patterns in head of, 130*f*

number of body segments in, 11*f*

Ubx gene expression reflecting maxilliped development in, 129*f*

Cryptic genetic variation, 188–190

Crystallins, 186

Ctenophores, 3, 106

D

Darwin, Charles, 97, 194

Deformed gene, 22

Degenerate polymerase chain reaction analysis, 100

Delta gene, 187

DER pathways, 189

Deuterostomes, 6, 7*f*

   engrailed genes of, evolution of, 104–105

Hox genes of, 110

phylogeny of, 99*f*

Development

genetic regulation of, 51–92

genetic toolkit for, 15–47

modularity of, 174

morphological diversity and, 13

Developmental regulatory genes, 97

conservation of, 98–101

history of, 108–112

Developmental regulatory hierarchies, 51

Devonian age, 112

Devonian tetrapods, 143

Diploblast phyla, 106

Diploblastic animals, 3

Dipteran

haltere of, conserved selector genes regulating, 145*f*

modern, 138*f*

Ubx regulation of hindwing development in, 141*f*

Disparity, body plans, 8

Distal-less (Dll) genes, 27, 44–46

in abdominal segments, 135–136

expression of, 69

   in arthropods and onychophora, 136*f*

   in butterfly eyespot foci, 156

   in insect abdominal limb evolution, 141

Diversification

animal design and, 8–12

of anteroposterior body organization in arthropods and vertebrates, 124–135

of body plans, 8

during Cambrian period, 2–4

combinatorial control in, 91–92

DNA and, 13, 173–194

of epithelial appendages, 151–152

evolution of, 1

of eyespot color patterns and regulatory gene expression domains, 158*f* of homologous body parts, 12f, 144–145

of insect wing morphology, 139–140

mechanisms of, 123–124

morphological, in conserved body plan, 135–144

selection and, 194

serially homologous structures and, 174

toolkit expansion and, 113

DMEF-2 genes, 32

DNA

cis-regulatory, 52–53, 175–181

diversity and, 13

gene structure at level of, 51–52

in regulatory evolution, 173–194

DNA-binding proteins, 176

activator, 179

frequency of various length sites of, 179t

probably frequencies of various length sites for, 178t

DNA-binding transcription factor, 26

DNA rearrangements, 108, 180

of cis-regulatory elements, 177f

DNA sequence databases, 100–101

Dorsoventral axis coordinate system, 68–71, 72

development of, 65

developmental genes for, 35

evolution of, 112

Dorsoventral patterning genes, 38f

Dorsoventral polarity, 42

Dorsoventral regulatory hierarchy, 84–87

dpp gene, 65

morphogene gradient, 75f

transforming growth factor-β, 37

Dragonfly, wing morphology of, 139f

Drosophila

embryogenesis of, 58–62

evolution of wings in, 137–139

gene complexes in, 22

gene divergence in, 105

genes in wing development of, 157

genetic toolkit of, 19–32

signaling pathways in, 41t

haltere, Ultrabithorax-regulated hierarchy in, 78f

homeotic mutants of, 20, 21f

Hox genes of, 23f, 43–46

homeodomains of, 27f

new functions for, 111–112

insect abdominal limb evolution in, 141–142

lab gene of, 183

life cycle of, 58f

limbless insect abdomen in, 135–136

segment polarity gene of, 125–126

systematic searches for developmental genes in, 32–33

trunk *Hox* gene expression and body architecture of, 127–129

Ubx expression in evolutionary morphologic changes in, 143f

Ubx gene expression in wing morphology of, 139–140

vestigial gene of, 183, 184f

wing-patterning genes of, 152

wings of, genetic regulatory hierarchy in, 73f

Drosophila melanogaster, 18

body plan of, 57–77

Duplications, 108

of cis-regulatory elements, 177f

DNA rearrangements and, 180

mapping of onto animal phylogeny, 104–105

relative timing of, 104

E

Ecdysozoan, 6, 7f

Hox genes of, 108–110

phylogeny of, 99f

Echinoderms

body plans of

evolution of, 166–167

radially symmetric, 168f

in early fossil record, 4f

structural novelties of, 167

Ediacaria

fauna, 3

forms, extinction of, 4

fossil, 5f

EGF-R, 41t

signal transduction pathways of, 189

Embryonic fields, patterning of, 47

Embryonic polarity, 33f

Embryos

Drosophila, 32f

organizers in, 16–18

En-1 protein, 84–85

Endoderm, 16

engrailed (en) gene, 28–31, 84

deuterostome, 103–104

expression of, 62

in butterfly eyespot evolution, 156

segmental, conservation of, 126f

regulatory activity of, 72

Epidermal growth factor, 37

Epithelial appendages

development of, 151f

diversity of, 151

functions of, 150–151

novel, 150–152

Evolution. *See also* Morphological evolution; Regulatory evolution

of cis-regulatory elements, 175–181

of genetic control of segmentation in arthropods, 125–127

genetic mechanisms underlying, 1
Hox domains shift during, 133–135
of morphological novelties, 149–169
of proteins, 97–98
of radical body plan change, 163–167
regulatory, of homologous body parts, 144–145
of vertebrate body plans, 125
Eyeless (ey) gene, 26–27, 28*f*, 29*f*, 45
Eyes
    evolution of, 117
    Pax6 expression during development of, 115*f*
Eyespots
    diversity of color patterns in, 157–158
    evolution of, 155–159
    specification of focus, 155–156

F

Feathers, 150–152
FGF8 expression, 84
FGF-R, 41*t*
Fibroblast growth factor signalling pathway, 37
Field-specific selector genes, 26–28, 29*f*, 44–46
Fins
    evolution of to limbs, 159–160, 191
    unpaired, 159
Fish
    paired pectoral and pelvic fins of, 142–143
    vertebrae of, 11*f*
Flight appendages, dorsal, 57
Forebrain, 162
    elaboration of, 163
Forelimbs
    loss of in snake, 165–166
    regulation of identity, 83, 89, 90*f*
    serially homologous, 125

tetrapod, morphologic changes in, 143
vertebrate, diversity of, 143–144
Forkhead (Fkh) transcription factor, 69–70
Fossil record, 2–5
    gaps in, 8
    Pre-Cambrian, 5*f*
    of snake hindlimb reduction and loss, 166
Fossilized insect forms, interpretation of, 152
Fossils, arthropod, 125
Frog
    homeotic transformations in, 20*f*
    vertebrae of, 11*f*
Fruit flies. *See Drosophila*

G

Gap genes, 34, 35
    encoding zinc finger proteins, 36–37
    sequential activation of, 65
    transcriptional activation of and cross-regulation by, 59, 60–61
Garcia-Bellido, Antonio, 51
GDF5 signaling protein, 89
Gene cluster, expansion of tandemly linked, 101*f*
Gene complexes, 22
Gene duplication, 101–104
    in gene divergence, 106*f*
    large-scale, 119
Gene families
    analysis of, 97
    history of, 98–108
Gene logic, 55
Gene trees, 101
General transcription factors, 52
Genes. *See also* Regulatory genes
    activation of, 55
        versus repression of, 181–182
    in arthropod segmentation, 125–127

battery of, regulated by dorsal protein, 66*f*
in body pattern diversity, 123–124
classification of by developmental function, 19
combinatorial control of, 64
concentration-dependent response of, 64
divergence of, 105
    mechanisms of, 106*f*
expression of
    logic and mechanisms controlling, 90–92
    methods for visualizing, 24*f*
    repression of, 175
    spatial boundaries of, 175
    specificity and diversity of, 91–92
    tissue-specific losses of, 186
loss of, 105
mapping, of onto animal phylogeny, 104–105
regulation of in metazoans, 51–53
repression of, 55
    versus activation of, 181–182
visualization of activity of, 53
Genetic redundancy, 43–44
Genetic regulatory hierarchy
    architecture of, 53–57
    dorsoventral, 66*f*
    in *Drosophila* wing, 73*f*
    of segmentation, 59*f*
Genetic regulatory logic, 56*f*
Genetic switches, 57
Genetic toolkit. *See* Toolkit genes
Genomic duplications, large-scale, 105
Germ layers, 16
Gould, Stephen Jay, 1
Gradients, 16
    activation of, 61*f*
    generation of, 59, 60
Grasshoppers, *Dll* and *Ubx* coexpression in pleuropod appendages of, 137*f*

H

Hagfish, neural crest derivatives of, 162
Hair, 150–152
Hairy gene, 36–37
Hassiophis terrasanctus, hindlimbs of, 166
Hedgehog gene, 37, 41*t*
Hedgehog pathway, 62, 157
Hedgehog signal family, 89
Helix-loop-helix motif, 36
Helix-loop-helix structure, 39*f*
    developmental function of, 40*t*
Helix-turn-helix motif, 36
Helix-turn-helix structure, 39*f*
HES1 gene, 80
Hierarchy, 57
    model of, 56*f*
Hindbrain
    Hox gene expression in, 81*f*
    Hox regulation in, 82
    vertebrate, rhombomeres of, 163
Hindlimbs
    dissociation of developmental programs of, 10
    reduction and loss of, 166
    regulation of identity, 83, 89, 90*f*
    serially homologous, 125
    tetrapod, morphologic changes in, 143
    vertebrate, diversity of, 144
Hindwings
    Ubx gene expression in morphologies of, 140*f*
    Ubx regulation of, 141*f*
Holometabolous insects, 57–58
Homeobox, 26
Homeodomains, 26, 39*f*
    developmental function of, 40*t*
    of *Drosophila Hox* genes, 27*f*
    in multicellular organisms, 106–107
Homeotic genes, 19–26
    systematic screening for, 21–22

Homeotic mutations, cell autonomy of, 22*f*
Homeotic selector genes, 35
Homeotic transformations, phenocopies of, 189
Homologous appendages, serially, 143–144
Homologous body parts
    divergence of through change in target genes, 145*f*
    diversification of, 10, 12*f*
    regulatory evolution and diversification of, 144–145
Homologous structures, diversification of, 124
Homologous trunk segments, 124
Homologs, 103
Housekeeping genes, 18, 98
Hox complex
    evolution of, 108–112
    expansion of, 108–110
    new functions for, 111–112
    sister complex to, 110–111
Hox genes, 22–25, 67–68
    arthropod head evolution and, 130–131
    in body plan diversity, 123–124
    cloning of, 22–25
    conservation of in vertebrates, 183*f*
    conserved between onychophora and arthropods, 129
    deployment of in limb field, 87–89
    domains of
        correlation of vertebrate axial patterning with, 131–133
        shifting of during evolution, 133–135
    of *Drosophila,* 23*f*
    in evolution of insect wings, 137–139
    expression and function of in vertebrate limb development, 88*f*

expression of
    annelid, 131
    in arthropod head, 130*f*
    modifications of in limb fields, 144
    modulations of in insect diversity, 140–142
    shifts in mirroring arthropod body changes, 127–129
    in fin evolution, 159–160
    in forelimb and hindlimb identity regulation, 89, 90*f*
    in identity versus formation of field, 26
    metazoan, evolution of, 109*f*
    mutations of, 89
    in novel epithelial appendage evolution, 152
    patterns of expression, 25–26
    regulation of expression, 67
    regulatory hierarchy modification in secondary fields by, 75–77
    sharing of among animals, 43–46
    in tetrapod axial identity, 132*f*
Hox ground plan, 67–68
    vertebrate, 80–82
Hox proteins
    binding sites for, 138–139
    functional conservation of, 113–114
    in organ primordia regulation, 70*f*
    in target gene activation, 178
Human forelimb, conserved selector genes regulating, 145*f*
hunchback gap gene, 61*f*, 126–127
Hunchback gradient, 60

I

Imaginal discs, larval, 25
Immunolocalization, 24*f*
Individualization, 8
    genetic mechanisms of, 10

Insects
    abdominal limbs of, 140–141
    body plan of, 57–58, 124
        dorsoventral axis coordinate
            system in, 65
        from egg to segments in,
            58–62
        Hox ground plan of, 67–68
        major features of, 57
        modified regulatory
            hierarchies in, 75–77
        organization of, 127–128
        patterning within, 71–74
        secondary fields in, 68–77
        segmentation genes in,
            62–65
        wing patterning regulation
            in, 74–75
    bristles of
        butterfly scales evolved
            from, 155f
        evolutionary relationship to
            butterfly scales, 154
    egg organizer of, 17f
    Hox genes of
        expression patterns in head
            of, 130f
        new functions for, 111–112
    limbless abdomen of, 135–136
    number of body segments in,
        11f
    Ubx gene expression in
        hindwing morphologies of,
        140f
    wings of
        diversification of morpholo-
            gy of, 139–140
        evolutionary origin of, 153f
        evolved from dorsal leg
            structures, 152–154
        number of, 136–139
In situ hybridization, 24f
Intercellular signals, 70f

J

Jacob, François, 149

K

Kangaroo, forelimb morphology
    in, 144
King, Mary-Claire, 123
Kreisler genes, 82
Krox20 genes, 82

L

Labial gene, 22
Lampreys, neural crest derivatives
    of, 162
Leech, Hox expression patterns
    in, 131f
Leg structures, insect wing
    evolution from, 152–154
Lepidoptera
    conserved selector genes
        regulating hindwing of, 145f
    modern, 138f
    Ubx regulation of hindwing
        development in, 141f
    wing scales of, 154
Leucine zipper motif, 36, 39f
Library screening, 100
Life cycle, toolkit genes in, 39–41
Limb buds
    AER formation and
        dorsoventral patterning of,
        86f
    organizer of, 16
    outgrowth of, 84–87
    positioning of, 83
        along rostrocaudal axis, 84
    regulation of formation, 85f
    temporal phases of Hox gene
        expression in, 88f
Limb fields, 69
    Hox genes deployment in,
        87–89
    position of, 84
Limb organizers, 17f
    establishment and signaling
        from, 83
Limb pattern elements,
    differentiation of, 89
Limb primordia, 69

Limbless insect abdomen,
    135–136
Limbless tetrapods, 150, 165–166
Limbs
    development of
        Hox gene in, 88f
        vertebrate, 82–84
        in vertebrates, 83f
    diversity of, 142–144
    dorsal patterning of, 84–85
    evolution of from fins,
        159–160
    morphology of
        divergence in functional
            evolution of, 82–83
        variation of in salamanders,
            192f
    pectoral, 159–160
    pelvic, 159–160
    reduction and elimination of in
        insects, 136–137
    vertebrate, 82–84, 125,
        142–144
Living fossils, 98–100
Lophotrochozoa, 6, 7f
    Hox genes of, 108–110
    phylogeny of, 99f

M

Mammals, evolution of body
    plans in, 125
Manx gene, reduced expression
    of, 164–165
Mass extinction, between
    Proterozoic and Cambrian, 4
Maternal effect genes, 32
    controlling embryonic polarity
        in Drosophila, 33f
Maternal transcription factor
    gradients, 60
    gene activation and, 61f
Maternal morphogens, 36f
Maxilliped development, 129f
Meristic differences, 11f
Meristic variation, 10
Mesoderm, 16
    lateral plate, 79

paraxial, 79
  segmentation of, 80
  vertebrate, 79–80
Messenger RNA (mRNA), 52
Metazoans
  anatomical and developmental features of, 6
  gene regulation in, 51–53
  Hox gene evolution of, 109*f*
  phylogeny of, 7*f*, 99*f*
  relationships of, 6
Mirror-image duplications, 28
Mirror-image polarity, 16
Modern animals, origins of, 3
Modular organization, 1, 8–10
  of body plans and body parts, 9*f*
Molecular biology Internet servers, 100–101
Molecular clocks, 3
Molecularly based phylogenies, 1
Molgula oculata, 164–165
Mollusca, early fossil record, 4*f*
Morgan, T.H., 15
Morphogen-producing organizers, 16–18
Morphogenetic field, 18
Morphogens, 16
  classical, 46–47
  maternal, 36*f*
Morphological complexity, 112–113
Morphological discontinuities, large-scale, 8–10
Morphological diversity, 1, 97–98, 124
  of animal body patterns, 123–124
  cis-regulatory DNA in, 186–190
  in conserved body plan, 135–144
  DNA and, 13
  evolution of regulatory DNA and, 190–194
  of insect wing, 139–140
Morphological evolution
  cryptic genetic variation and potential for, 188–190

for older structures, 150–159
  regulatory evolution primacy in, 173–174
  trends in, 8
Morphological novelties
  definition of, 150
  evolution of, 149–169
  from older structures, 150–159
Morphological traits, artificial selection for, 187–188
Multibranched appendages, dorsal branch of, 153*f*
Multicellular embryo, organized, 112
Multicellular life, 3
Muscle precursors, 70–71
  rates of, 182
Myrapods, 124
  body organizations of, 127–128

N

Nanos gene, 33*f*, 35, 61*f*
National Center for Biotechnology Information (NCBI), 100
Nautilus gene, 32
Necessity, 55
Nematodes
  in early fossil record, 4*f*
  loss of toolkit genes in, 113
Network, regulatory, 56–57
Neural crest
  cell proliferation of, 162
  evolution of, 161–163
Neural precursors, 70–71
Neurogenic genes, 35
Nodal points, regulatory information flowing through, 53–55
Notch gene, 37
Notch signaling, 151
Notochord
  evolution of, 163
  loss of, 163–165
Novelty. *See* Morphological novelties; Structural novelties

evolution of, 10
  in vertebrates, 159–163
  regulatory evolution in, 174
  origin of, 167–169
  time scale of, 169
Nymph
  primitive mayfly, 138*f*
  primitive winged, 138*f*

O

Olenoides serratus, 2*f*
Onychophorans, 8
  body plans of
    diversification in, 123
    gene expression in, 129
  diversity in, 124
  Dll and *Ubx* expression in, 136*f*
  number of body segments in, 8–10
Oral-aboral axis, 166–167
Ordovician fish, 159
Organ primordia, 70*f*
Organizers, 16
  classical, 46–47
  morphogen-producing, 16–18
Organizing signals
  integration of to form proximodistal axis, 87–89
  in secondary field patterning, 71–74
  in wing patterning, 74
Origin of Species, 194
Orthologs, 103
Outgroup, 98–100

P

Pachyrhachis problematicus, hindlimbs of, 166
Pair-rule genes, 34
  initiation of periodic expression of, 61–62
  primary, 61
  segmentation, 36–37, 125–126
  sequential activation of, 65
  stripes of, 62

regulation of, 63*f*

transcriptional regulation of, 60

Pair-rule proteins, regulating segment polarity gene expression, 60, 62

Paired appendages, evolution of, 159–160

ParaHox genes, 110–111

complex of, 112

Paralogous genes, evolution of, 105

Paralogs, 103

Parasegment, 26

Pax6 gene, 45

expression of during eye development, 115*f*

orthologs, 114–115

small eye gene, 46*f*

Pax transcription factors, 107–108

pdm genes, 152–154

Pectoral limbs, *Hox* genes in evolution of, 159–160

Phenocopies, 189

Phenotypic diversity, 188–190

DNA level and, 173

Phyla, comparative anatomy of, 1

PHYLIP, 101

Phylogenetic inferences, 98–101, 114–117

Phylogenetic relationships, 1, 6–8

Phylogeny

of deuterostome *engrailed* genes, 103*f*

mapping gene duplication events relative to, 97

metazoan, 99*f*

Pikaia gracilens, 2*f*

Platyhelminths, early fossil record, 4*f*

Pleiotropy, 39–43, 174

Pleuropod appendages, *Ubx-Dll* coexpression in, 137*f*

Polychaetes, 131*f*

Polymerase chain reaction, degenerate, 100

Porifera, 106

in early fossil record, 4*f*

Precambrian animal fossils, 3, 5*f*

Precis coenia hindwings, 156*f*

Priapula, 4*f*

Primary fields, 18

formation of, 57

Primitive bilaterian body plan, complexity of, 117

Primitive winged adult insect, 138*f*

Primordia, 18

secondary field development from, 71–74

Proboscipedia gene, 22

Promoters, 52

Proteins

domains of in multicellular organisms, 106–107

evolutionary changes in, 97–98

Protostomes, 6, 7*f*

Proximodistal axis, 87–89

Pteridinium, fossil, 5*f*

Pupal stage, 57

Python vertebrae, 11*f*, 133

Q

Quantitative morphological variation, 150

Quantitative trait loci (QTLs), 187

Quantitative traits, 187

R

Radiation, bilaterian, 119

Radical fringe expression, 84

Redundancy, genetic, 105

Region-specific selectors, 26

Regulatory circuits, 117

Regulatory DNA, 190–194

Regulatory evolution

in animal body pattern diversity, 123–124

characteristics of, 174

of homologous body parts, 144–145

in morphological evolution, 173–174

origin of novelties and, 167–169

primacy of, 173–194

Regulatory genes

conservation of, 98–101

in insect body pattern organization, 59

in vertebrate limb diversity, 142–144

Regulatory hierarchy, 65

gene expression patterns and, 92

Hox genes modification of, 75–77

models of, 57

Regulatory linkages, 167–168

Regulatory logic, 55–57

Regulatory networks, 89

Regulatory proteins, 51–53

Repeated parts, changing number of, 10

Reporter genes, 53

analysis of, 54*f*

Rhombomeres, 80

hindbrain segmentation into, 81*f*

RNA polymerase II, 52

Rostrocaudal axis, 84

S

Salamanders, 192*f*

Salivary gland primordia, 69–70

scabrous gene, 187

Scales, 150–152

butterfly wing, 154, 155*f*

Scalloped (sd) gene, 27, 45

Secondary fields, 18

formation of, 57

Hox gene modification of regulatory hierarchy in, 75–77

integrating anteroposterior and dorsoventral coordinate systems, 68–71

patterning within, 71–74

Segment polarity genes, 35, 125–126

in *Drosophila* segments, 64*f*

regulation of, 60, 62
sequential activation of, 65
Segment polarity proteins, 60
Segmental identity, 19–26
Segmentation, 57
in anteroposterior axis, 79–80
evolution of genetic control of in arthropods, 125–127
genetic regulatory hierarchy of, 59f
of hindbrain, 81f, 82
insect and crustacean head, 130–131
of paraxial mesoderm, 80
regulatory connections in cascade of, 126–127
Segmentation genes
expression of, 37f
lessons from, 62–65
mutants of, 34f
Segmented body, development of, 58–62
Selection, variation effects of, 194
Selector genes, 21
in forelimb and hindlimb identity regulation, 89, 90f
homologous structure divergence and, 145f
in secondary field patterning, 71–74
in vertebrate limb diversity, 142–144
Sensory Mother Cell (SMC), 154
Serially homologous structures
diversity of, 10, 181–182
morphological diversification of, 174
Sevenless tyrosine kinase receptor, 189
Sex combs reduced (Scr) gene, 22, 26
Shh protein, restriction of, 85–86
Signal integration, 72–74
Signal transduction pathways, 91
Signaling pathways, 46–47
conservation of, 113–114
in *Drosophila* genetic toolkit, 41t

gene families of, 107t
pleiotropic components of, 42–43
in segment polarity gene regulation, 62
transduction, 42f
Signaling proteins, 46–47
in combinatorial regulation of wing patterning, 74
Small eye gene, 45
snail gene, 65
Snakes
evolutionary transitions in limblessness of, 167f
leg loss of, 165–166
morphological novelty of, 150
Somites, 79
in anteroposterior axis, 79–80
diversification of, 181
Spatial patterns
combinatorial control of, 91–92
progressive development of, 90–91
Specificity, combinatorial control in, 91–92
Spemann, Hans, 51
Spemann organizers, 16, 17f, 46
Sponges, 3, 7f, 106
Stabilizing selection, 184–186
Stickleback fish body patterns, 193f
Structural novelties, 149
evolution of, 149–169
Syncytium, 60
Systematic mutagenesis screens, 32–33

T

T-box genes, 89, 90f
T-box transcription factor, 144
T gene, 163, 164f
Tail, loss of, 163–165
Tandemly linked gene clusters, 101–102
Target genes
activation of, 178
battery of, 57

Dpp morphogen gradient in activating, 75f
evolutionary change in regulation of, 158–159
Tbx5 gene, 89
Telencephalon
elaboration of, 163
expansion of cortical region of, 162
Terminal differentiation genes, 175
Tertiary fields, 18
Tetraploidization, 102–103, 105
Tetrapods
appendages of, 159
autopod patterning structures of, 168
body organization of, 124
evolutionary adaptation of, 125
hand of, 8
evolution of, 159–160
origin of digits, 161f
Hox genes in evolution of axial identities in, 132f
limbless, 165–166
limbs of
cis-regulatory elements in evolution of, 160
distal element of, 160
morphologic changes in, 143
Thermoregulation, epithelial appendages in, 150–151
Tinman (tin) genes, 27–28, 44–46
Toll gene, 41t
Toll signalling pathway, 37
Toolkit genes
analysis of, 100–101
in animal design, 47
assembly of, 105–108
cis-regulatory elements of, 182–183
critical features of, 18–19
for development, 15–47, 117–119
Drosophila, 19–32
evolution of, 97–119

expansion of in morphological complexity, 112–113

expression of, 35–36

function of, 91

identification of in different animals, 98–101

interpreting, 112–117

loss of in nematodes, 113

minimal, 107

pleiotropy of, 39–43, 174

products of, 36–38

sharing of among animals, 43–47

Trace fossils, 3

trans-acting regulatory elements, 65–67

Transcription factors, 36–37

acting on *Hox* genes, 67

binding sites for, 175, 176–178

conservation of, 185*f*

in *cis*-regulatory system, 175

DNA-binding, 179

in *Drosophila* genetic toolkit for development, 40–41*t*

functional conservation of, 113–114

genes in, 107*t*

structural motifs of, 39*f*

toolkit gene encoding, 18

Transcriptional activators, 52–53

of gap genes, 59

Transcriptional regulator binding sites, 183–184

Trichome patterning, 141–142

Trilobite, 11*f*

Triploblastic, 3

Trunk *Hox* gene expression, 127–129

Twist (twi) gene, 32, 36–37, 65

U

Ultrabithorax (Ubx) genes, 20, 21*f*, 22, 67, 189

cis-acting regulatory elements of, 68*f*

expression of, 25

with abdominal limbs in butterfly larvae, 142*f*

anterior boundaries of, 129

in arthropods and onychophora, 136*f*

in *Drosophila* evolutionary changes, 143*f*

in *Drosophila* wing morphology, 139–140

in maxilliped development in crustaceans, 129*f*

regulation of insect hindwing development, 141*f*

Ultrabithorax (Ubx) protein

expression of and arthropod body architecture, 127–128

regulated hierarchy, 78*f*

selective gene expression and, 77

Unsegmented animals, *Hox* genes in, 44

Urbilateria

development of, 114–117

rebuilding of, 116*f*

V

Vascular system, water, 167

Vertebrae

changing number of, 131–133

variation in numbers of, 8–10

Vertebrates

axial morphology of

correlation of with *Hox* expression domains, 131–133

Hox genes regulating, 45*f*

body plans of

anteroposterior axis in, 79–80

development of, 77–89

evolution of, 125

forelimb and hindlimb regulation in, 89

Hox ground plan in, 80–82

limb development in, 82–84

organizing signals in proximodistal axis and deploying *Hox* genes in, 87–89

outgrowth of limb bud in, 84–87

positioning limb buds along rostrocaudal axis, 84

regulatory networks in, 89

brain of

elaboration of, 162*f*

evolution of, 161–163

subdivisions of, 162–163

diversity of anterior/posterior body organization in, 124–135

downstream regulatory changes in limb diversity of, 142–144

embryo segmentation and somite formation in, 79*f*

evolution of novelties of, 159–163

eye development in, *Pax-6/small eye* gene controlling, 46*f*

forelimbs of

dissociation of developmental programs, 10

diversification of, 12*f*

homeotic transformations in, 20*f*

Hox gene complexes and expression in, 44*f*

large-scale genome duplications in, 102*f*, 112

meristic differences among, 11*f*

radiation of, 98–100

serially homologous appendages of, 143–144

Vestigial (vg) gene, 27, 29*f*, 72–75

W

Waptia fieldensis, 2*f*

Wilson, Allan, 123

Wing primordia, 69

Wingless gene, 37, 41*t*

Wingless gene, 62

Wingless insect, 138*f*

Wings. *See also* Insects, wings of

    combinatorial regulation of
      patterning, 74

    evolution of number of,
      136–139

Wnt-7a protein, 85

Wnt signaling, 42–43, 62, 151

Z

zen gene, 65

Zinc fingers

    developmental function of,
      40*t,* 41*t*

    encoding of, 36–37

    motif of, 36, 39*f*

Zone of polarizing activity (ZPA),
    16, 17*f,* 46–47, 166

    in *Hox* gene expression, 87

    restriction of, 85–86

Zuckerkandl, Emile, 173

Zygote, 32

Zygotic dorsoventral patterning
    genes, 65

Zygotic genes, 38*f*